About the Authors

RICHARD SHENKMAN is the editor of George Mason University's History News Network (www.hnn.us) and can be seen regularly on CNN and Fox News. He is the author of five history books including the *New York Times* bestseller *Legends, Lies, and Cherished Myths of American History*. His most recent book is *Presidential Ambition: Gaining Power at Any Cost*.

KURT REIGER holds degrees from Vanderbilt University and the Wharton School of the University of Pennsylvania. He lives in Oklahoma City and heads a company that specializes in innovative horticulture and that can be found on the Web at www.treebag.com.

ONE-NIGHT STANDS

WITH

AMERICAN HISTORY

★ ≋≋

Odd, Amusing, and
Little-Known Incidents
(Revised Edition)

RICHARD SHENKMAN AND
KURT REIGER

Perennial
An Imprint of HarperCollins*Publishers*

Grateful acknowledgment is made for permission to quote from the following:

From *An Informal History of Texas* by Frank X. Tolbert. Copyright © 1951, 1961 by Frank X. Tolbert Sr. Reprinted by permission of Harper & Row, Publishers.

From Allan Nevins, *The Gateway to History* (Garden City: Anchor, 1962), p. 151. Reprinted by permission of the Trustees of Columbia University in the City of New York as copyright owner.

From *Huey Long* by T. Harry Williams. Copyright © 1970 by T. Harry Williams. Reprinted by permission of Alfred A. Knopf, Inc.

From *The Anatomy of the Anecdote* by Louis Brownlow. Copyright © 1960 by Louis Brownlow. Reprinted by permission of the University of Chicago Press.

From *Webster's New International Dictionary*, Second Edition. Copyright © 1959 by G. & C. Merriam Company. Reprinted by permission of G. & C. Merriam Company.

A previous edition of this book was published in 1980 by HarperCollins Publishers.

HarperCollins books may be purchased for educational, business, or sales promotional use. For information please write: Special Markets Department, HarperCollins Publishers Inc., 10 East 53rd Street, New York, NY 10022.

First Perennial edition published 2003.

Designed by Nancy Singer Olaguera

Library of Congress Cataloging-in-Publication Data

Shenkman, Richard.
    One-night stands with American history : odd, amusing, and
little-known incidents / Richard Shenkman and Kurt Reiger.—
Rev. ed., 1st Perennial ed.
        p. cm.
    Includes bibliographical references and index.
    ISBN 0-06-053820-1
    1. United States—History—Anecdotes. 2. United States—
History—Humor. I. Reiger, Kurt. II. Title.

E178.6.S473 2003
973'.02'07—dc21   2003042957

03 04 05 06 07 ❖/RRD 10 9 8 7 6 5 4 3 2 1

THIS EDITION IS DEDICATED TO

OUR PARENTS;

CRISTI, HALLIE BRUGGE, AND ELLIOTT FREI REIGER;

AND

JOHN STUCKY

CONTENTS

★ ≋

Preface

★ ≈≈≈

Everybody likes a good story or an interesting odd fact. This is a book of them—stories about inventions, presidential secrets, hoaxes, rare incidents, and cultural idiosyncrasies. It is not a book of tall tales in the genre of Paul Bunyan or Johnny Appleseed. The great majority of the stories are not well known and will surprise even professional historians. We found them by searching through biographies and histories and by asking specialists in American history for suggestions. While all of the odd facts are true, some of the anecdotes are questionable, but we have included only those which strike a chord of verisimilitude. A few of them are clearly apocryphal, such as a story by Parson Weems, but they are included here because they are characteristically American.

ACKNOWLEDGMENTS

★ ≋

This book could not have been written without the help of those people who unselfishly gave us their personal support as we traveled around the country doing research: Richard Grosse, Wendy Enikeieff, Leigh Thalhimer, Tom Neylan, Kyle Bogertman, Wes Jenkins, David McNatt, Lillian Thomas, Mark Ordan, Jane Hertzmark, Robert Rutledge, Chuck Covington, George Slover, Ellen Rodgers, Melanie and Tex Young, and Tom Reagan.

Many librarians aided us in our research. In particular, we would like to thank the staff of Vanderbilt University's Joint University Library and Bernice Laks and the reference librarians at Vassar College.

Three people deserve special acknowledgment: M. Glen Johnson, Holman Hamilton, and Dan Kilgore. Others who suggested stories that appear in this book include: Dan Aaron, Thomas Bailey, Robert Barr, John Barsotti, Robert Becker, Hugh Bell, Frank Bergon, Shirley Bill, Wallace Bishop, John Boles, Paul Boller, Jr., Paul Boyer, Clayton Brown, Mark T. Carleton, Thomas Charlton, John Clark, Robert Collins, Virginius Dabney, Hugh Davis, John Davis, Ronald Davis, L. G. DePauw, Douglas Eckberg, Clifford Egan, Tuffly Ellis, Donald Fleming, Ben Ford, George Forgie, William M. Fowler, Joe B. Frantz, Steve Gerstel, Arrell M. Gibson, Steve Gietschier, Kenneth Goldstein, Francis Gosling, Lewis Gould, Dewey Grantham, Donald Green, Robert A. Gross, Jack Haddick, Hal Halliday, Dagmar S. Hamilton, David Helms (the famous tree farmer from Arlington, Washington), John Higham, P. P. Hill, Michael Holt, Kathy Jacob, Kent Keith, Mary-Jo Klein, Lawrence Knutson, William Leuchtenburg, David Levy, John Lindsay, Arthur Link, John L. Loos, Roger W. Lotchin, Peyton McCrary, Catherine McDowell, James McIntosh, Jack Maguire, Alexander Marchant, Robert Merrick, Howard Merriman,

James W. Mooney, James T. Moore, H. Wayne Morgan, John M. Murrin, Norman Pollack, Stuart Postle, George E. Reedy, Robert Remini, Don Ritchie, Ray Robinson, Richard C. Rohrs, Philip J. Schwarz, Edward S. Shapiro, William F. Sharp, Randi Shenkman, Larry Shore, Henry J. Silverman, Elbert B. Smith, Stuart Sprague, Lenny Steinhorn, Peter Stillman, Ann Thomas, Charles Vandersee, Frank E. Vandiver, Ben Wall, David J. Weber, Russell Weigley, Bernard A. Weisberger, Richard White, Donald Worcester, Donald Yoder.

One-Night Stands

with

American History

★ ≋

BEGINNINGS

★ ≋

Puritanism: "The haunting fear that someone, somewhere, may be happy."

—H. L. MENCKEN

★ SCRAPBOOK OF THE TIMES ★

- In 1631, Massachusetts Bay outlawed chimneys made of wood after major fires ravaged several towns.
- Concerned about the fluctuating value of money, Willem Kiefft, deputy-general of New Amsterdam, issued an order in the 1640s that wampum be strung tightly together. Loose wampum had created problems of exchange and led to an increase in bartering.
- The first income tax in American history was imposed in 1643 by the colonists of New Plymouth, Massachusetts.
- Wall Street received its name in 1644, when New York City built a wall around lower Manhattan to protect cattle from marauding Indians.
- The first person convicted and executed in America for witchcraft was Margaret James of Charlestown, Massachusetts. She was executed on June 15, 1648, almost fifty years before the notorious trials at Salem.
- When inflation became a major problem in the 1650s in New Amsterdam, Peter Stuyvesant, head of the colony, imposed price controls—at first just on bread, brandy, and wine, and later on shoes, stockings, soap, salad oil, candles, vinegar, and nails.
- The first Bible was printed in America in 1661—in the Algonquin language, a language that no one today can read.
- In the seventeenth century New Englanders spoke with a Southern accent. The Southern accent was a survival from old England and predominated in America until the eighteenth

century, when Yankees began speaking with the familiar twang.

- For wearing silk clothes, which were above their station, thirty young men were arrested in 1675 in New England. Thirty-eight women were arrested for the same offense in Connecticut.

- To celebrate the end of King Philip's War, the worst Indian war in their history, New England colonists, on August 17, 1676, placed the head of the man who started it, Chief Metacomet, on a pole outside the gates of Plymouth. The head remained there for twenty-five years.

- The rhymes of Mother Goose, a real person, were first published in 1719 under the title *Songs for the Nursery; or, Mother Goose's Melodies for Children*. Her son-in-law, a printer, who was annoyed by the rhymes Mother Goose sang to his baby, published them in an attempt to embarrass her.

- Women were in such short supply in Louisiana in 1721 that the government of France shipped twenty-five prostitutes to the colony. By this action the government hoped to lure Canadian settlers away from Indian mistresses.

- Angered by the poor quality of dormitory food, students at Harvard College rebelled in 1766. The administration responded by suspending half the student body.

★ COLUMBUS'S SECRET LOG ★

On September 9, 1492, as the last land dropped below the horizon, Christopher Columbus began keeping two logs. One log, which he kept secret, was a true reckoning of his course and distance. The other was a falsified account of the ship's location written so the crew would not be frightened at sailing so far from land. Yet as fate would have it, Columbus overestimated his distance by 9 percent in his private log, placing his discovery much farther west than it actually was. The false log, however, contained no such "error." Columbus had given his sailors a record that was, for all practical purposes, virtually correct.

SOURCE: Samuel E. Morison, *Christopher Columbus* (Boston: Mentor, 1955), p. 36.

★ TOBACCO: SIXTEENTH-CENTURY PANACEA ★

A relationship between smoking and health was recognized soon after the introduction of tobacco to Europeans. In 1588, Thomas Hariot published *A Brief and True Report of the New Found Land of Virginia*, in which he described the new product to the Old World. "It openeth all the pores and passages of the body," he wrote. Users "are notably [preserved] in health, and know not many greevous diseases wherewithall wee in England are oftentimes afflicted."

SOURCE: Thomas Hariot, *A Brief and True Report of the New Found Land of Virginia* (1588; rpt. Ann Arbor, Mich.: Edwards Brothers, 1931), p. 63.

★ A NAMING IN THE NEW WORLD ★

Plymouth, Massachusetts, was named by the Pilgrims in 1620 because the *Mayflower* had sailed from Plymouth, England. It sounds logical and is believed by most people, but it isn't true. In 1614, Captain John Smith sailed from Jamestown, Virginia, on his first exploring mission to the northeast. He returned with a map cluttered with "barbarous" names representing Indian villages. Smith showed the map to Prince Charles and asked His Royal Highness to provide good English names in place of the Indian ones. Prince Charles obliged, and changed the Indian name of Accomack to Plymouth, years before any white man ever settled there.

SOURCE: *All the People Some of the Time* (Ann Arbor, Mich.: William L. Clements Library, 1941), p. 8.

★ THE PILGRIMS DIDN'T LAND ON THE ROCK ★

The belief that the Pilgrims landed on Plymouth Rock rests solely on the recollection of a ninety-five-year-old man 120 years after the event. In 1741, Elder Thomas Faunce told a crowd that his father, who arrived in America three years after the *Mayflower*, had once pointed out to him the rock as the place where the Pilgrims had landed. There is no other evidence for the tradition.

Besides, the Harvard historian Edward Channing proved that the ship never could have landed at the rock given the direction of the current.

SOURCES: Arthur Lord, *Plymouth and the Pilgrims* (Boston: Houghton Mifflin, 1920), pp. 120–21; Samuel E. Morison, *By Land and by Sea* (New York: Knopf, 1953), p. 307.

★ REDS IN PLYMOUTH ★

When the Pilgrims arrived in America in 1620, they immediately committed an un-American act—at least, one that would be so viewed later on. Desiring to create a just and equal society, they established a communist economy. The early colonists remained committed to communism for several years, until they finally decided that it was inefficient. Their switch to capitalism was a defeat of sorts, since it implied the inability of men to work hard for the common good without individual incentive.

SOURCE: William Bradford, *Of Plymouth Plantation*, ed. Harvey Wish (New York: Capricorn Books, 1962), p. 90.

★ BOY EXECUTED FOR BUGGERY ★

Strange as it may sound, in 1642 the Pilgrim colony at Plymouth, Massachusetts, was struck by a crime wave. William Bradford, governor of the colony, described in his history of Plymouth one of the worst crimes:

"Ther was a youth whose name was Thomas Granger; he was servant to an honest man of Duxbery, being aboute 16. or 17. years of age. (His father & mother lived at the same time at Sityate.) He was this year detected of buggery (and indicted for the same) with a mare, a cowe, tow goats, five sheep, 2. calves, and a turkey. Horrible it is to mention, but the truth of the historie requires it. He was first discovered by one that accidentally saw his lewd practise towards the mare. (I forbear perticulers.) Being upon it examined and committed, in the end he not only confest the fact with that beast at that time, but sundrie times before, and at severall times with all the rest of the forenamed in his indictmente. . . . And accordingly he was cast by the jury, and condemned, and after executed the 8. of September, 1642. A very sade spectakle it was; for first the mare, and then the cowe, and the rest of the lesser catle, were kild before his face, according to the law, Levit: 20:15, and then he him selfe was executed."

SOURCE: William Bradford, *Of Plymouth Plantation*, ed. Harvey Wish (New York: Capricorn Books, 1962), pp. 202–3.

★ PURITANS PROHIBIT CHURCH WEDDINGS ★

In 1647 the New England Puritans did something which might seem odd in view of their professed and very real piety: they outlawed the preaching of wedding sermons. Even before that year they had mandated that all marriage ceremonies be conducted by a civil magistrate.

Why? The Puritans believed that marriage was a fundamentally secular institution, of no direct concern to the church. It was, as Martin Luther wrote, not a sacrament, but "a secular and outward thing, having to do with wife and children, house and home, and with other matters that belong to the realm of government, all of which have been completely subjected to reason." By the end of the century the Puritans relaxed their restrictions on church involvement in weddings and allowed marriage ceremonies to be performed by ministers as well as by justices of the peace.

SOURCE: Daniel Boorstin, *The Americans: The Democratic Experience* (New York: Random House, 1973), p. 67.

★ THE PURITANS WHO STOLE CHRISTMAS ★

The Puritans have been blamed for nearly everything that is wrong with America, but they cannot be blamed for the commercialization of Christmas. In colonial Massachusetts it was illegal to observe Christmas. By a law passed in 1659, anybody "found observing, by abstinence from labor, feasting or any other way, any such days as Christmas day" was fined five shillings for each offense. The law was repealed in 1681, but only because the Puritans were sure no one would celebrate the holiday. In 1685, Judge Samuel Sewall noted in his famous diary that on Christmas everyone went to work as usual. Not until the middle of the nineteenth century did Christmas become a major holiday.

SOURCE: Daniel Boorstin, *The Americans: The Democratic Experience* (New York: Random House, 1973), p. 158.

★ COLONIAL REGARD FOR THE LEGAL PROFESSION ★

Hostility to lawyers is not a recent phenomenon. It goes back at least to the Middle Ages and was widespread in seventeenth-century America. In 1641, Massachusetts Bay actually adopted a law making it illegal to earn money by representing a person in court; the law stayed on the books for seven years. In Virginia legislators went even further. In 1658 they passed a law expelling all attorneys from the colony. Not until 1680 was the law repealed and the lawyers allowed to return.

SOURCE: Mary Cable, *American Manners and Morals* (New York: American Heritage Publishing Company, 1969), pp. 26–27.

★ MARRIAGES IN THE NUDE ★

In colonial New England there were many instances of women getting married in the nude or in their underwear. Why? According to an old English tradition, if a woman married "in her shift on the king's highway," her husband would not be responsible for her prenuptial debts. To preserve decency, these marriages were often performed at night—but not always. There were some couples who found they could comply with tradition and still be discreet during daylight. In one case a couple got married while the woman stood naked in a closet, with only her hand showing. It is not known whether or not creditors generally accepted the tradition.

SOURCE: Alice Morse Earle, *Customs and Fashions in Old New England* (New York: Scribner's, 1893), pp. 77–79.

★ THE FOURTH COMMANDMENT IN ★ COLONIAL CONNECTICUT

In colonial Connecticut there were stiff legal penalties against disobedient children. Examples:

- "If any Childe or Children above *fifteen years old*, and of sufficient understanding, shall Curse or Smite their natural Father or Mother, he or they shall be put to death, unless it can be sufficiently testified, that the Parents have been unchristianly negligent in the education of such Children."
- "If any man have a stubborn or rebellious Son, of sufficient understanding and years, *viz. fifteen years of age*, which will not obey the voice of his Father, or the voice of his Mother, and that when they have chastened him, he will not hearken unto them; then may his Father or Mother, being his natural Parents, lay hold on him, and bring him to the Magistrates assembled in Court, and testifie unto them, that their Son is Stubborn and Rebellious, and will not obey their voice and chastisement, but lives in sundry notorious Crimes, such a Son shall be put to death, *Deut.* 21: 20–21."

SOURCE: George Brinley, *The Laws of Connecticut* (Hartford, Conn.: privately printed, 1865), pp. 9–10.

★ KING JAMES PRESERVES COURT ETIQUETTE ★

The religious tension between egalitarian-minded Quakers and the Stuart monarchy which had driven William Penn from England during the reign of Charles II was evident in an encounter Penn had with Charles's brother, James II. Penn had always been on good terms with James, and visited him soon after the king's coronation. Upon entering the monarch's presence, Penn failed to remove his hat. James immediately removed his.

"Friend James," inquired Penn, "why dost thee uncover thy head?"

"Because," replied the new king, "it is the fashion here for only one man to wear his hat."

SOURCE: *The Journal of Solomon Sidesplitter* (Philadelphia: Pickwick and Company, 1884), p. 124.

★ EARLY AMERICAN JUSTICE ★

In 1691 boat trader John Clark was found dead. His stolen supplies were uncovered in the home of Thomas Lutherland, an indentured servant from New Jersey. Lutherland was immediately arrested on a charge of murder.

At the trial Clark's body was brought forward. To prove Lutherland's guilt or innocence, the court ordered the defendant to touch the corpse. The verdict would be based on a superstition widely believed in the New World that a dead body would bleed if touched by its murderer. Lutherland placed his hand on his postmortem accuser, but the cold body remained the same.

Unfortunately for Lutherland, the court was not entirely bound by the confines of superstition. The defendant was found guilty anyway and executed on February 23, 1691.

SOURCE: Jay Robert Nash, *Bloodletters and Badmen* (New York: M. Evans and Company, 1973), p. 345.

★ WITCHES ON LSD? ★

The "witches" of Salem, Massachusetts, who in 1692 swore they had seen "the devil at work," may have been simply a group of young girls hallucinating from contaminated bread. Researchers have recently postulated that the "witches" were suffering from ergotism, a toxic condition produced by eating grain tainted with the parasitic fungus ergot, genus Claviceps. Ergot is a hallucinogen related to lysergic acid diethylamide, or LSD. Nineteen men and women were found guilty and hanged as a result of the trial, while one man was pressed to death between two stones.

SOURCE: *New York Times*, March 31, 1976, p. 1.

★ THE SCARLET LETTER ★

In 1695 the Puritans in Salem, Massachusetts, passed the law against adultery that suggested to Nathaniel Hawthorne the story of *The Scarlet Letter*. The law provided that people convicted of adultery would have to wear the letter "A" on a conspicuous part of their clothes for the remainder of their lives. The law also made adulterers liable to a severe whipping of forty lashes and required them to sit in humiliation on the gallows with chains about their necks for at least one hour. Harsh as these penalties were, however, they were much milder than the punishments common in New England just a few years before. In the middle of the seventeenth century the penalty for adultery in Massachusetts was death.

In just one year during the third quarter of the seventeenth century, when the population of Boston was only 4,000, there were forty-eight instances of bastardy and fifty of fornication.

SOURCES: Joseph B. Felt, *The Annals of Salem* (Salem, Mass.: privately printed, 1827), p. 317; Mary Cable, *American Manners and Morals* (New York: American Heritage Publishing Company, 1969), p. 28.

★ THE MOST PROLIFIC PURITAN ★

Even for a Puritan, Cotton Mather showed extraordinary industriousness. The youngest student ever admitted to Harvard College, he published more than 450 books and pamphlets during his life. The works included histories, biographies, essays, sermons, and fables and concerned theology, philosophy, science, and medicine. Critics have praised the works generously.

Cotton Mather's father, Increase, also wrote many books, but did not come close to his son's record. Increase wrote just 130 books, though he contributed to more than sixty-five others. Samuel Mather, Cotton's son, was even less productive. He wrote a mere twenty volumes.

SOURCE: James D. Hart, *The Oxford Companion to American Literature* (New York: Oxford University Press, 1965), pp. 531–34.

★ A LOCAL SOLUTION FOR A LOCAL PROBLEM ★

The colonial towns of Lyme and New-London, Connecticut, once held conflicting claims to the same piece of land. Its value, at that time, was regarded as a trifling amount—certainly not enough to warrant the appointment of representatives from the two towns to present their cases before the colonial legislature. Instead, the towns agreed on a local solution. Champions were selected—Griswold and Ely for Lyme, and Ricket and Latimer for New-London—and on the appointed day these four met on a designated field and slugged it out with their fists. Griswold and Ely beat up Ricket and Latimer, and Lyme took possession of the disputed tract.

SOURCE: Leonard Deming, *A Collection of Useful, Interesting, and Remarkable Events* (Middlebury, Conn.: J. W. Copeland, 1825), p. 314.

★ THE TRANSVESTITE GOVERNOR ★

There once was a governor of New York who was a transvestite. His name was Lord Cornbury, and he served from 1702 to 1708. A favorite of the Queen, he appeared at public ceremonies in full drag, wearing a dress, silk stockings, and an elaborate hairdo. He let his nails grow long and customarily donned high-heeled boots. He remained as governor for six years until the American colonists, outraged by his behavior, finally forced his recall.

Portrait of Lord Cornbury by an unidentified artist. (The New-York Historical Society.)

★ THE SLAVES GET THEIR DAY ★

In the eighteenth century the slave was hardly better off than he was in the nineteenth century. But there were times when it seemed that he was. When the governor of Virginia wanted to make sure that his slaves did not get drunk on the Queen's Birthday in 1711, he had to strike a bargain with them. In exchange for their good behavior on that day, he promised to allow them to become as drunk as they wished the following day. The bargain worked well for both parties: the governor had well-behaved servants on the Queen's Birthday; the slaves enjoyed an extraordinary feast the next day. In the nineteenth century the master would probably never have bargained with his slaves; he would simply have ordered them to do as he wished.

The slaves who worked for William Byrd of Westover, one of Virginia's most prominent landowners, were not half as fortunate as those who labored for the governor. Byrd often whipped his slaves, many times for no good reason. Once he whipped a slave to punish his wife. She had whipped one of his slaves after he warned her not to. So he retaliated by whipping one of her slaves. In his diary, where he tells about this incident, he does not say if his whipping had a chastening effect on his wife.

SOURCES: Edmund S. Morgan, *Victorians at Home* (Williamsburg, Va.: Colonial Williamsburg, Inc., 1952), p. 67; Louis B. Wright and Marion Tinling, eds., *The Secret Diary of William Byrd of Westover, 1709–1712* (Richmond, Va.: Dietz Press, 1941), p. 533.

★ FUNERAL GIFTS ★

In the seventeenth and eighteenth centuries it was customary to provide guests at funerals with gifts, including a black scarf, a pair of black gloves, and a mourning ring. Eventually people accumulated large collections of these items. One Boston minister noted that he possessed several hundred rings and pairs of black gloves. Even people who did not attend a particular funeral were sometimes sent one or more of these symbols of death. When Judge Samuel Sewall refused to go to the funeral of the notoriously wicked John Ive, he still received a pair of gloves. "I staid at home," he wrote in his diary, "and by that means lost a Ring." In 1721 laws began to be passed limiting the gifts to pallbearers and clergymen. During the Revolution the custom of giving scarves and gloves was abandoned, since the items could no longer be imported. Instead, people began using black armbands as a sign of mourning.

SOURCE: Mary Cable, *American Manners and Morals* (New York: American Heritage Publishing Company, 1969), p. 45.

★ FRANKLIN'S MAGAZINE ★

Ben Franklin was certain he had hit upon another great idea. He would publish the first magazine in America and make it as much a success as England's *Gentleman's Magazine* and the *London Magazine*. But he could not do all of the work himself, so he hired John Webbe, a contributor to his newspaper, to do the heavy work. Unfortunately, Webbe was not an honest man. Before Ben even wrote one article, Webbe secretly went to a printer and arranged for the publication of his own magazine. On February 13, 1741, the first issue of Webbe's *American Magazine* went on sale. Just three days later, the *General Magazine and Historical Chronicle, for all the British Plantations in America,* published by Franklin, made its debut.

SOURCE: Frank Luther Mott, *A History of American Magazines* (New York: D. Appleton, 1930), I, 73–74.

★ CLASS STATUS AT COLONIAL HARVARD ★ AND YALE

According to Samuel Eliot Morison, who wrote a three-volume history of Harvard, before the early 1700s the college probably ranked students on the basis of an educated guess as to their academic performance. Afterward, at the insistence of Brahmin parents, the college adopted a policy of using family status to define class rank. The *jeunesse dorée* of New England, therefore, were automatically put at the top of the class, leading the way in academic processions, prayer, and other functions. As the size of both the college and the community increased, however, the new ranking system proved too unwieldy for the "egalitarian" New Englanders. The college abandoned the practice in 1769, when a man named Phillips protested loudly about the ranking of a boy ahead of his own whose father had not been a justice of the peace as long as the elder Phillips. About the same time Yale, which had also ranked students according to family status, adopted an alphabetical arrangement.

At the College of William and Mary, second oldest college in the United States, the question of "placing" students by any other criterion than performance did not arise. As Jane Carson, research assistant at Colonial Williamsburg, explained, "Where status is clearly understood, there is no need for formal labels."

SOURCE: Gerald Carson, *The Polite Americans* (New York: William Morrow, 1966), p. 33.

★ NO DAY FOR A HANGING ★

In 1768 the French colony of New Orleans rebelled against its new Spanish owners. By July 24, 1769, when the Spanish army arrived to put things back in order, the rebellion had faltered. Its leaders then fell over one another attempting to be first in line to explain away their regrettable actions and to promise unceasing loyalty to Spain. A trial ensued, and on October 24 five of the defendants were sentenced "to the ordinary pain of the gallows." The verdict was to be carried out the next day.

But the official hangman of New Orleans in 1769 was a black man. Spanish officials felt that under the circumstances a white man should conduct the executions. After all, these were former leaders of the French colony, not simply common criminals. A reward was offered, but by the next day no whites had volunteered to act as hangman. Spanish leaders, however, did not want to postpone the execution. So, rather than have a black man hang the traitors, they substituted a firing squad. At three o'clock on October 25 the five condemned men were shot in the New Orleans public square. The Spanish commander, Alejandro O'Reilly, has since been known in Louisiana history as "Bloody O'Reilly."

SOURCE: Herbert E. Bolton, *The Spanish Borderlands* (New Haven: Yale University Press, 1921), p. 248.

★ POOR RICHARD DISCUSSES DISEASES ★

"This Year the Stone-blind shall see but very little; the Deaf shall hear but poorly; and the Dumb shan't speak very plain. And it's much, if my Dame *Bridget* talks at all this Year. Whole Flocks, Herds and Droves of Sheep, Swine and Oxen, Cocks and Hens, Ducks and Drakes, Geese and Ganders shall go to pot; but the Mortality will not be altogether so great among Cats, Dogs and Horses. As for old Age, 'twill be incurable this Year, because of the Years past. And towards the Fall some people will be seiz'd with an unaccountable Inclination to roast and eat their own Ears: Should this be call'd Madness, Doctors? I think not.—But the worst Disease of all will be a certain most horrid, dreadful, malignant, catching, perverse and odious Malady, almost epidemical, insomuch that many shall run Mad upon it; I quake for very Fear when I think on't; for I assure you very few will escape this Disease; which is called by the learned Albumazar *Lacko'mony*."

SOURCE: Van Wyck Brooks, ed., *Poor Richard* (New York: Paddington Press, 1976), p. 73.

★ AS POOR RICHARD SAYS . . . ★

Some striking quotations from Benjamin Franklin's *Poor Richard's Almanack:*

- "He that lies down with Dogs, shall rise up with fleas."
- "Men & Melons are hard to know."
- "God works wonders now & then; Behold! a Lawyer, an honest Man!"

- "Three may keep a Secret, if two of them are dead."
- "Fish & Visitors stink in 3 days."
- "He that lives upon Hope, dies farting."
- "Keep your eyes wide open before marriage, half shut afterwards."

SOURCE: Van Wyck Brooks, ed., *Poor Richard* (New York: Paddington Press, 1976), *passim*.

★ FRANKLIN ADVOCATES DRINKING MADEIRA ★

"'Friend Franklin,' said Myers Fisher, a celebrated Quaker lawyer of Philadelpia, one day, 'thee knows almost everything; can thee tell me how I am to preserve my small-beer in the back-yard? my neighbours are often tapping it of nights.'

"'Put a barrel of old Madeira by the side of it,' replied the doctor; 'let them but get a taste of the Madeira, and I'll engage they will never trouble thy small-beer any more.'"

SOURCE: Reuben Percy, ed., *The Percy Anecdotes* (London and New York: Frederick Warne & Company, [1887]), IV, 124.

INVENTING A COUNTRY

★ ≈≈≈

"A monarchy is like a merchantman. You get on board and ride the wind and tide in safety and elation but, by and by, you strike a reef and go down. But democracy is like a raft. You never sink, but, damn it, your feet are always in the water."

—MASSACHUSETTS FEDERALIST FISHER AMES

★ SCRAPBOOK OF THE TIMES ★

- When Phillis Wheatley's *Poems on Various Subjects*, the first book ever published by an American Negro, appeared in London in 1773, someone bound a copy of the book in Negro skin.
- The Boston Tea Party fixed in the American mind the belief that the British tax on tea was an appalling burden. But it was not. Tea actually cost less in America than in Britain—even with the tax.
- Despite the important matters it had to consider, the Continental Congress often wasted hours discussing trivial concerns. Once it debated whether one James Whitehead ought to receive $64 in compensation for feeding British prisoners (they paid him). Another time they argued about the case of a wagonmaster who wanted $222.60 for transporting goods for the army to Dobbs Ferry and Cambridge. After much debate they finally agreed to pay him, too.
- The best-selling book in 1783 was Noah Webster's *American Spelling Book*.
- Thomas Jefferson once described the White House as "a great stone house, big enough for two emperors, one pope and the grand lama in the bargain."
- The median age of Americans in 1800 was sixteen.

- When Aaron Burr killed Alexander Hamilton in 1804, Burr was vice president of the United States.
- In 1805 half the Harvard student body was suspended after rioting against the poor quality of dormitory food.
- In 1815, Napoleon Bonaparte's brother, Joseph, exiled king of Spain, arrived in America and established a home on a 211-acre estate near Bordentown, New Jersey. He lived there for seventeen years before returning to Europe.
- The word "buncombe" and its derivative, "bunk," both meaning "speechmaking to please constituents" or "nonsense," were coined during the debate over the Missouri Compromise of 1820, when Felix Walker, congressman from Buncombe County, North Carolina, stood up in the House and said he wanted "to make a speech for Buncombe." The speech was irrelevant and rambling and transformed the name of the county he represented into a word.
- On the day Thomas Jefferson died, friends were soliciting money for his relief at a ceremony in the House of Representatives marking the fiftieth anniversary of the Declaration of Independence. The former president's assets had dwindled considerably and he desperately needed cash. Had he lived, however, he would not have been able to depend on this solicitation. According to John Quincy Adams, only four or five people at the ceremony contributed to Jefferson's relief.
- James Madison's last words were: "I always talk better lying down."
- In 1820, when Captain John White, the first American to make contact with Vietnam, sailed into Saigon Harbor the Vietnamese asked him for uniforms and guns.
- The worst day in the history of the New York Stock Exchange was March 16, 1830, when a mere thirty-one shares, valued at $3,470.25, were traded.

★ THE BATTLE OF BUNKER HILL ★

The Battle of Bunker Hill was the first major conflict of the American Revolution. Unfortunately, it did not occur on Bunker

Hill. The revolutionary Committee of Safety ordered colonial officers to seize and fortify Bunker Hill against possible attack from British regiments attempting to control Boston Harbor. But colonial military leaders, for reasons still unknown, entrenched instead on Breed's Hill, a smaller mound some two thousand feet away. The famous battle occurred on that hill, on June 17, 1775, but popularly became known as the Battle of Bunker Hill.

Today, Breed's Hill, where the battle took place, is known as Bunker Hill. The original Bunker Hill is covered with houses.

SOURCE: George W. Stimpson, *Nuggets of Knowledge* (New York: A. L. Burt, 1934), p. 162.

★ REVOLUTIONARY AID FROM A TRANSVESTITE ★

In 1775, Arthur Lee, a colonial agent for Massachusetts, met French playwright Caron de Beaumarchais in England. With the American Revolution already in its infant stages, Lee lost no time trying to persuade Beaumarchais that the French government should strike at its old enemy, the English, by aiding the colonies. When Beaumarchais returned home, he carried Lee's message to the French court. Spurred on by his efforts, the French government supplied approximately 90 percent of the munitions used by the colonists in the first two years of war. Eventually, of course, the French entered the war on the side of the Americans. Almost every contemporary observer and historian agrees that without the aid of the French the American Revolution would have taken a very different course.

But why was Beaumarchais in England? The man who was instrumental in persuading the French to help the colonies was traveling on a secret mission to retrieve stolen documents from the Chevalier d'Éon, a transvestite, about whose sex no one was certain. D'Éon was a championship fencer—in the female competition. Once a captain of the Grenadiers proposed to d'Éon. Some contemporaries even reported that Beaumarchais believed d'Éon was in love with the playwright. On the other hand, d'Éon was also a captain of the French dragoons and was a former diplomatic agent—both male occupations. At d'Éon's death in 1810, over thirty-five years later, an autopsy found "the male organs of generation perfectly formed in every respect."

Sources: Samuel Flagg Bemis, *A Diplomatic History of the United States*, 3d ed. (New York: Holt, 1950), p. 19n; Louis de Loménie, *Beaumarchais and His Time* (New York: Harper & Brothers, 1857), p. 226.

★ FRANKLIN PREDICTS A HANGING ★

When the Declaration of Independence was adopted, the chances that the Revolution would actually succeed were slim. At the signing ceremony John Hancock remarked, "We must be unanimous—we must all hang together." "We must indeed all hang together," agreed Ben Franklin, "or, most assuredly, we shall all hang separately."

SOURCE: Daniel George, ed., *A Book of Anecdotes* (n.p.: Hulton Press, 1957), p. 147.

★ AMERICAN INDEPENDENCE NOT DECLARED ★ ON JULY FOURTH

The second of July and not the fourth should be celebrated as the anniversary of American independence. It is true that the Declaration of Independence was dated "July 4, 1776," but independence itself had been declared two days earlier. All that happened on the fourth was the approval of the final draft of the document in a vote that was not even unanimous, despite the claim made in the opening of the declaration that it was; New York did not agree to the declaration until July 19. The signing of the document did not take place on the fourth, though many people believe it did. John Hancock and the secretary of the Congress did sign one copy of the declaration that day, but the official signing ceremony occurred on the second of August, with six members signing later, one not until 1781.

On July 3, 1776, John Adams predicted in a letter to his wife that "the Second day of July, 1776, will be the most memorable Epocha, in the History of America. I am apt to believe that it will be celebrated, by succeeding Generations, as the great anniversary

The signing of the Declaration of Independence. (*Harper's Weekly*, July 3, 1858, p. 417.)

Festival." When this letter was published in the nineteenth century, an editor changed the date of the letter to July 5, and had Adams advising his wife that "the Fourth day of July, 1776" would be honored as the anniversary of U.S. independence.

SOURCE: Catherine Drinker Bowen, *John Adams* (Boston: Little, Brown, 1950), p. 598n.

★ THE LIBERTY BELL HOAX ★

Until George Lippard, a Philadelphia journalist, immortalized the Liberty Bell in 1847 in his *Legends of the American Revolution*, Americans did not care about the bell. Lippard, a latter-day Parson Weems, invented the whole story about the bell ringing in American independence. The only thing true about his story was that the bell did hang in the Philadelphia statehouse in 1776 when the Founding Fathers drafted the Declaration of Independence. But no one thought of ringing it. In 1828 the city of

Philadelphia tried to sell the bell as scrap, but could find no buyers: the bell simply was not worth the expense of removing it from the building. The first time anyone referred to the bell as the Liberty Bell was in a pamphlet entitled "The Liberty Bell, by Friends of Freedom," distributed at the Massachusetts Anti-Slavery Fair in 1839. In the pamphlet the bell symbolized the freedom of black slaves, not the independence of white Americans from Britain.

SOURCES: Daniel Boorstin, *The Americans: The National Experience* (New York: Random House, 1965), pp. 381–82; *American Heritage*, June 1973, p. 104.

★ THE SUBMARINE: AN AMERICAN INVENTION ★

The first war submarine in history was not built by the Germans in World War I, but by the Americans in the Revolution. The vessel, dubbed the *Turtle*, was the invention of David Bushnell, a Yale graduate. It consisted of a wood frame, several small windows, and a hand-driven propeller. Big enough for only a single person, it reached speeds of up to three miles an hour and could stay submerged for approximately thirty minutes. On the outside was an egg-shaped time bomb equipped with an iron screw for penetrating the hulls of enemy ships.

The *Turtle* was used only once during the Revolution. In August 1776, operated by First Sergeant Ezra Lee, it was launched in New York Harbor to attack the British *Eagle*, a sixty-four-gun warship commanded by Admiral Howe. Unfortunately, though Lee successfully maneuvered the submarine to the side of the man-of-war, he was unable to attach the bomb to the ship's copper-sheathed frame. The failure doomed the prospects of submarine warfare. Although George Washington was enthusiastic about submarines and had even helped finance the *Turtle*, the Continental Congress refused to provide any funding.

SOURCES: John Oliver, *History of American Technology* (New York: Ronald Press, 1956), pp. 100–101; Roger Burlingame, *March of the Iron Men* (New York: Scribner's, 1938), pp. 146–48.

★ SECRET HISTORY OF THE BATTLE OF TRENTON ★

On the day after Christmas, 1776, a loyalist spy appeared at the headquarters of Hessian commander Colonel Johann Rall with an urgent message. The spy had learned that George Washington and his small Continental army had secretly crossed the Delaware River that morning and were advancing on Trenton, where the Hessians were encamped. He attempted to enter Rall's headquarters, but was stopped and told to write down his message. The colonel had left strict orders that no one was to disturb his liquor or cards.

A porter took the message into the house and handed it to the Hessian colonel. But rather than interrupt his deal, Rall thrust the note, unread, into his pocket. By the time his deal was over, the message that could have had an important effect on the course of the Revolution had been completely forgotten.

The colonel would regret his orders. He was still playing cards when the guards of the camp began discharging their muskets in a futile attempt to stop Washington's army. The patriots' attack had come as a complete surprise. Without time to organize or rally, the entire Hessian army was captured. The colonists had gained their first major victory of the Revolutionary War.

During the battle Colonel Rall fell mortally wounded. As he lay dying, he swore that if he had read the loyalist's message the revolutionaries would never have taken either his army or his life.

SOURCE: George W. Stimpson, *Nuggets of Knowledge* (New York: A. L. Burt, 1934), p. 141.

★ A BRITISH SOLDIER DECIDES NOT TO ★ KILL WASHINGTON

Several days before the Battle of Brandywine Creek, in September 1777, George Washington and a French officer reconnoitered the area of land and water that stretched between the American camp at Chadd's Ford and the British camp at Kennett Square, four miles away. George Washington always preferred doing his own reconnaissance, especially in a case such as this, where only a few

paltry maps of the region were available. Of course, it was dangerous for the commander of the American army to take to unprotected fields, but Washington was the kind of man who willingly
took risks when he had to. At times he was downright reckless. A
few days before, on another reconnoitering mission, he slept in a
house that a friend suspected was filled with British sympathizers.

As he traveled about, Washington seemed completely oblivi-
ous to the threat of attack. He took few precautions and galloped
everywhere. At one point, following the Frenchman, he rode into
a clearing in the woods—where he would be an easy target for
even the worst rifleman.

Unfortunately, hiding nearby was a band of four British sharpshooters who had thrown themselves to the ground at the sound
of approaching horses. The leader of the soldiers was one Patrick
Ferguson, a master marksman, who had invented a deadly accurate rifle that weighed only seven and a half pounds. Ferguson was
on his very first campaign, having convinced the government,
after an amazing performance before King George III, of the usefulness of sharpshooters.

When the Frenchman and Washington rode up, Ferguson saw
a golden opportunity to prove the worth of his outfit. Obviously
the pair were important. One was wearing hussar, the other a buff
and blue uniform and "a remarkable large cocked hat," as Ferguson
later noted. Instantly Ferguson notified his men "to steal near to
them and fire at them." But as his men readied, Ferguson peremp-
torily withdrew his order. He had suddenly decided it would be
better to capture the stately pair than to kill them. Without hesitating, he shouted out to the Frenchman, who was nearer, ordering him to dismount. The Frenchman ignored the command,
however, and called out a warning to his friend in buff and blue.
Washington promptly wheeled his horse around and made off,
with the Frenchman close behind. The two men then sped back to
the American lines, safe and unharmed.

During the Battle of Brandywine, Ferguson was wounded in
the elbow of his shooting arm and sent to a hospital. There he told
of his encounter with the two enemy officers and aroused curiosity
as to the identity of the escaped men, whom he described in

detail. One morning, as Ferguson subsequently recalled, a surgeon, "who had been dressing the wounded rebel Officers, came in and told us that they had been informing him that General Washington was all that day with the light troops, and only attended by a French Officer in Hussar dress, he himself dressed and mounted in every point as described."

Ferguson afterwards remarked, "As I was within that distance at which in the quickest firing, I could have lodged half a dozen of balls in or about him before he was out of my reach, I had only to determine; but it was not pleasant to fire at the back of an unoffending individual who was acquitting himself coolly of his duty, and so I left him alone."

In 1779, Ferguson lost his life in a battle he might well have prevented from ever taking place had he, on September 7, 1777, killed the commander of the American army.

SOURCE: Reginald Hargreaves, "The Man Who Almost Shot Washington," *American Heritage*, December 1955, pp. 62–65.

★ FRAUDULENT BIOGRAPHY OF JOHN PAUL JONES ★

"[In 1900, A. C. Buell published] a two-volume life of John Paul Jones which was praised by the *American Historical Review*, accepted as an authority in various universities, and recommended to students of the Naval Academy at Annapolis. It seemed sober history; it was actually a mixture of authentic and manufactured materials. When his sources ran thin, Mr. Buell calmly manufactured new ones. He invented a French memoir upon Jones by Adrien de Cappelle, a volume of papers by the North Carolina worthy Joseph Hewes, and a printed French collection of Jones's own papers, drawing 'facts' liberally from these imagined storehouses. He invented collections of papers by Robert Morris and Gouverneur Morris in places where they had never existed, and so obtained more 'facts.' He invented a will by William Jones of North Carolina in order to give John Paul an estate, and had Jones deposit 900 guineas in the Bank of North America in 1776—a wonderful feat, for the Bank was not established until 1781. In short, as Albert Bushnell Hart has written, this inventor of materials recalls Mark Twain's praise of the duckbilled platypus, so versatile and gay. 'If he wanted eggs,' remarked Mark Twain, 'he laid them.'"

SOURCE: Allan Nevins, *The Gateway to History*, rev. ed. (Garden City, N.Y.: Anchor, 1962), p. 151. Reprinted by permission of the Trustees of Columbia University in the City of New York as copyright owner.

★ ETHAN ALLEN AT CHURCH ★

"Allen was in church one Sunday with a number of friends listening to a very high Calvinistic minister (exact stature not recorded). The text chosen was, 'Many shall strive to enter in, but shall not be able,' and the preacher premised his remarks by observing that the grace of God was certainly sufficient to include one person out of ten. 'Secondly' disclosed the fact that not one in twenty would attempt to avail himself of salvation. At 'thirdly' it came out that but one man in fifty was really an object of Divine solicitude. 'Fourthly' was announced and the estimate of elect now reduced to great correctness, the sad conclusion being drawn that but one of eighty—when Allen [famous as an atheist] seized his hat and evacuated the pew, 'I'm off, boys; any one of you may take my chance.'"

SOURCE: *Harper's Magazine*, July 1875, p. 308.

★ AMERICA'S FIRST PRESIDENT ★

The first president of the United States was not George Washington. In 1781, Maryland finally signed the Articles of Confederation, and the union among the thirteen states became an actuality. John Hanson, the man who signed for Maryland, was immediately elected president of the assembly. His formal title was "President of the United States in Congress Assembled."

Even George Washington himself addressed Hanson as "President of the United States." When Washington won his great victory at Yorktown, Hanson sent the general a letter of congratulation.

Washington reciprocated at once, addressing his letter to the "President of the United States."

SOURCE: David Wallechinsky and Irving Wallace, *The People's Almanac* (Garden City, N.Y.: Doubleday, 1975), p. 261.

★ THE BEST BIRD FOR THE JOB ★

The bald eagle became America's national symbol when it was placed on the Great Seal in 1782. One member of Congress who did not support the winning selection was Benjamin Franklin. The eagle, he claimed, was too common a bird to be a national symbol. Later, he elaborated on his views in a letter to his daughter:

"For my part, I wish the bald eagle had not been chosen as the representative of our country; he is a bird of bad moral character; he does not get his living honestly; you may have seen him perched on some dead tree, where, too lazy to fish for himself, he watches the labor of the fishing hawk [waiting to steal that bird's food]."

Franklin thought a more uniquely American bird should have been selected by the Continental Congress. His choice, the turkey.

SOURCE: George W. Stimpson, *Nuggets of Knowledge* (New York: A. L. Burt, 1934), p. 315.

★ WOMAN SUFFRAGE IN THE ★
EIGHTEENTH CENTURY

The passage of the Nineteenth Amendment in 1920 gave every American woman the right to vote. Until then women were allowed to vote in only a dozen states. But the amendment was not as revolutionary as it seemed. Women in New Jersey had been granted the right to vote as early as 1776. At that time a new constitution was adopted which gave the suffrage to any free person worth more than fifty pounds. If a woman met the financial qualification, she could vote. The men who framed the constitution had not expected women to take advantage of the vote and were not trying to make the state more democratic. But their constitution inadvertently did open up the system to women—at least women who possessed more than fifty pounds.

At first few women availed themselves of the opportunity to cast a ballot. The constitutional loophole seemed harmless enough. So harmless, as a matter of fact, that it was retained when a new constitution was written in 1797. But the next few years saw women deciding closely contested elections. In 1807 the New Jersey legislature rescinded woman suffrage.

For sixty-one years woman suffrage lay dead. Then suddenly, in Wyoming Territory, the corpse sprang to life. Without controversy a measure granting women full rights passed the upper house of the legislature. But in the lower house the bill faced stern opposition. Men there were hard and down-to-earth, opposed to the high sentiments of the "uppers." They ridiculed the bill, added outrageous amendments, and considered not voting on it until July 4, 1870—when the legislature would no longer be in session. Finally, however, they passed the bill, by the not overwhelming majority of six to four. Of course, the members did not really want women voting. But since they were all Democrats, they decided it would be politically advantageous to let the governor, a Republican, who was known to oppose woman suffrage, veto the bill. He, then, and not they, would be blamed for defeating women's rights.

But Republican governor John Campbell would not be done in so easily. Though a young man and new to the state, he had a hound dog's nose for a dirty plot. On December 10, 1869, he signed the suffrage bill.

SOURCES: Kirk H. Porter, *A History of Suffrage in the United States* (Chicago: University of Chicago Press, 1918), pp. 136, 254; Lynne Cheney, "It All Began in Wyoming," *American Heritage*, April 1973, pp. 62–65.

★ WASHINGTON COLLAGE ★

- George Washington introduced the jackass to America.
- Washington's family motto was "*Exitus acta probat*" (the end justifies the means).
- When he died, Washington provided in his will for the eman-

cipation of his slaves on the death of Martha, his wife. Washington was the only member of the Virginia dynasty to free all of his slaves.

- Washington was one of the richest men in America. At his death his holdings were worth about half a million dollars and included: 33,000 acres of land in Virginia, Kentucky, Maryland, New York, Pennsylvania, Washington, D.C., and the Northwest Territory; $25,000 worth of stocks; 640 sheep; 329 cows; 42 mules; and 20 workhorses.

★ THINGS NAMED AFTER GEORGE WASHINGTON ★

1 state

7 mountains

8 streams

10 lakes

33 counties

9 colleges

121 towns and villages

SOURCE: Marcus Cunliffe, *George Washington* (New York: Mentor, 1958), p. 16.

★ MAKING IT ILLEGAL TO ATTACK THE ★ UNITED STATES WITH MORE THAN THREE THOUSAND TROOPS

At the Constitutional Convention a member moved that "the standing army be restricted to no more than five thousand men." When George Washington heard this, he turned to a friend and remarked that the resolution was fine with him—so long as the convention agreed to an amendment prohibiting armies from invading the United States with more than three thousand troops.

SOURCE: Paul Wilstach, *Patriots Off Their Pedestals* (Indianapolis: Bobbs-Merrill, 1927), p. 29.

★ MYSTERY AT WASHINGTON'S INAUGURATION ★

The image of George Washington standing erect, hand on Bible, as he recites the oath of office at his first inauguration at Federal Hall in New York City is a familiar one. But what was the weather like at the inauguration? Was it a clear day or rainy? The simple answer is that no one knows. Despite the importance of Washington's inauguration and public interest in the event, not one person at the time bothered to record the weather. Until the nineteenth century, oddly enough, Americans almost never commented on the weather when describing public events.

The first mention of the weather in New York on April 30, 1789, did not come until sixty-five years later, when Rufus Griswold, who wasn't even born until after the event, published Washington Irving's reminiscence of the historic occasion. According to Griswold, Irving remembered in 1854 that at eight o'clock on the morning of the inauguration the skies had been cloudy, but that by nine, when the ceremony began, the sun shone brightly.

Besides Irving, only one other eyewitness ever specifically noted the inaugural weather: Mary Hunt Palmer, the daughter of a prominent general, who in 1789 was fourteen years old. She provided her reminiscences at age eighty-three in 1858.

Mary Palmer's story was startlingly different from Irving's. She too remembered clouds, but her clouds were larger, darker, and more forbidding. Moreover, her clouds never disappeared, but brought down on the city a drenching rain. "It never rained faster," she recalled, "than it did that day." The rain was so heavy, actually, that George Washington had to carry an umbrella as he proceeded up the street to be sworn in.

Who is to be believed? Historically speaking, no one knows.

SOURCE: Charles Warren, *Odd Byways in American History* (Cambridge, Mass.: Harvard University Press, 1942), pp. 92–101.

★ WASHINGTON PAPERS GIVEN AWAY ★

Today a single letter written by George Washington can bring several thousand dollars. But when he died, just about anyone could obtain a sample of Washington's correspondence. The President's nephew, Bushrod Washington, who had control of the papers, simply could not resist giving them away. This tradition was carried on by Jared Sparks, the first editor of a published version of the papers. His twelve volumes were only a version because Sparks edited out of the papers all words or sentiments that did not seem in keeping with the great man. Though Sparks had only been given temporary custody of the letters, he did not hesitate to mutilate them, to snip a sentence here and a signature there to satisfy the request of some autograph hunter. In 1861, Sparks confessed to Richard Henry Dana Jr., that he could not furnish Dana with an autographed letter of the first president. "I have had many such," he wrote, "but the collectors have long ago exhausted my stock. The best I can do is to enclose a very small specimen of his handwriting."

SOURCE: Ralph K. Andrist, ed., *The Founding Fathers: George Washington* (New York: Newsweek, 1972), pp. 6–7.

★ PARSON WEEMS TELLS A STORY ABOUT ★ GEORGE WASHINGTON

"One day [George Washington's father] went into the garden, and prepared a little bed of finely pulverized earth, on which he wrote George's name at full, in large letters—then strewing in plenty of cabbage seed, he covered them up, and smoothed all over nicely with the roller.—This bed he purposely prepared close alongside of a gooseberry walk, which happening at this time to be well hung with ripe fruit, knew would be honoured with George's visits pretty regularly every day. Not many mornings had passed away before in came George, with eyes wild rolling, and his little cheeks ready to burst with *great news*.

"'O Pa! come here! come here!'

"'What's the matter, my son? what's the matter?'

"'O come here, I tell you, Pa: come here! and I'll shew you such a sight as you never saw in all your life time.'

"The old gentleman suspecting what George would be at, gave him his hand, which he seized with great eagerness, and tugging him along through the garden, led him point blank to the bed whereon was inscribed, in large letters, and in all the freshness of newly sprung plants, the full name of

GEORGE WASHINGTON

"'There, Pa?' said George, quite in an ecstasy of astonishment, 'did you ever see such a sight in all your life time?'

"'Why it seems like a curious affair, sure enough, George!'

"'But, Pa, who did make it there? who did make it there?'

"'It grew there by *chance*, I suppose, my son.'

"'By *chance*, Pa! O no! no! it never did grow there by *chance*, Pa. Indeed that it never did!'

"'High! why not, my son?'

"'Why, Pa, did you ever see any body's name in a plant bed before?'

"'Well, but George, such a thing might happen, though you never saw it before.'

"'Yes, Pa; but I did never see the little plants grow up so as to make one single letter of my name before. Now, how could they grow up so as to make *all* the letters of my name! and then standing one after another, to spell *my name* so *exactly!*—and all so neat and even too, at top and bottom!! Oh Pa, you must not say *chance* did all this. Indeed *somebody* did it; and I dare say now, Pa, *you* did it just to scare *me*, because I am your little boy.'

"His father smiled; and said, 'Well George, you have guessed right. I indeed *did* it; but not to *scare* you, my son; but to learn you a great thing I wish you to understand. I want, my son, to introduce you to your *true* Father.'"

SOURCE: Mason L. Weems, *The Life of George Washington* (Philadelphia: M. Carey & Son, 1818), pp. 15–16.

★ JOHN ADAMS NOT THE VICE PRESIDENT OF ★ A CRICKET CLUB

The first session of the United States Senate, which met from April 23 to May 14, 1789, consumed virtually all of its time in a drawn-out debate over what to call the president of the United States. John Adams, vice president, feared that unless specific words of dignity and preeminence were used in the title of the president, the nation's chief executive might be mistaken as the president of a fire company or perhaps a cricket club. Personally, Adams favored referring to the nation's leader as "His Highness the President of the United States and Protector of the Rights of the Same." The Senate finally decided on the title "His Highness," which was the way Parliament addressed the king of England. The House of Representatives, however, adopted the simple practice of calling the chief executive "The President of the United States," a title that quickly won the support of everyone. Senators themselves wanted to be called "The Honorable," but the House did not go along with that, either. One title was agreed to by both houses. Vice President Adams, by 1789, was quite overweight. Both senators and representatives agreed—privately—that he should be addressed as "His Rotundity."

SOURCE: John C. Miller, *The Federalist Era: 1789–1801* (New York: Harper & Row, 1960), p. 9.

★ THE AMERICAN FLAG ONCE HAD ★ FIFTEEN STRIPES

The American flag has not always had thirteen stripes. When Vermont and Kentucky came into the Union in the 1790s, Congress adopted a flag of fifteen stars and fifteen stripes. In 1818, Congress, not wanting to crowd the flag, voted to indicate the admission of new states by the addition of stars only. The Congress also voted to revert to a flag of thirteen stripes.

The flag of fifteen stars and fifteen stripes, adopted May 1, 1795. (D. Peleg Harrison, *The Stars and Stripes and Other American Flags* [Boston: Little, Brown, 1906], p. 65.)

★ THE ISLAND OF AMERICA ★

The Dey of Algiers did not have a good understanding of American geography or democracy. In one treaty he described the United States as an island "belonging to the islands of the ocean." In a letter to James Madison he insisted on addressing the President as "His Majesty, the Emperor of America, its adjacent and dependent provinces and coasts and wherever his government may extend, our noble friend, the support of the Kings of the nation of Jesus, the most glorious amongst the princes, elected among many lords and nobles, the happy, the great, the amiable, James Madison, Emperor of America."

SOURCE: Charles Warren, *Odd Byways in American History* (Cambridge, Mass.: Harvard University Press, 1942), p. 4.

★ THE PRICE OF PEACE ★

For many years Americans took pride in Charles Cotesworth Pinckney's boast, "Millions for defense, but not one cent for tribute." Historically, however, there wasn't a halfpenny's worth of truth to Pinckney's statement. In the early years of the Republic the United States paid tribute regularly to other countries to avoid war. In 1786 the United States gave $10,000 to Morocco, in 1795 it began "donating" an annual tribute of $21,600 to Algiers. In 1797 a treaty negotiated with Tripoli included among its promises American gifts of:

$40,000 in gold and silver coins

5 rings (3 with diamonds, 1 with a sapphire, 1 with a watch)

141 ells of fine cloth

4 caftans of brocade

$12,000 in Spanish currency

In 1805 the U.S. government paid Tripoli $60,000 for the return of captured American citizens.

The record of paying tribute to foreign governments was dismal, but the flip side of the problem was even worse. The same countries that demanded tribute also frequently offered presents of their own. The Constitution specifically prohibits any government official from receiving foreign gifts without the express authorization of Congress. This put American ministers and even the president in an awkward position. They could not refuse the present without insulting the giver, but they could not accept it without first asking the permission of Congress, which had passed a resolution against receiving gifts.

In 1806, Thomas Jefferson was presented with four Arabian horses by the Bey of Tunis. After much thought, he decided to accept the horses and then sell them, to defray the expense of maintaining the minister from Tunisia in a Washington hotel. The horses, however, turned out to be practically worthless. While the cost of the minister's stay came to over $15,000, the horses did not fetch more than fifty dollars apiece.

Similar problems resulted when other gifts were offered to the United States. These gifts included:

4 Arabian horses from Turkey (1832)

1 lion and 2 studs from Morocco (1835)

2 gold-mounted swords from Siam (1836)

2 lions and 2 horses from Morocco (1839)

2 Arabian horses, 1 string of 150 pearls, 2 large-size pearls, 1 carpet, 1 bottle of oil of rose, 4 cashmere shawls, 5 demijohns of rosewater, 1 gold-mounted sword from Oman (1840)

2 horses from Oman (1844)

In 1861 the king of Siam offered the United States dozens of elephants, but Lincoln politely refused them.

SOURCE: Charles Warren, *Odd Byways in American History* (Cambridge, Mass.: Harvard University Press, 1942), pp. 3–29.

★ Georgia Sells Alabama and Mississippi ★

Late in 1794 word leaked out that the Spanish government had finally come to terms. At the negotiating table the Spaniards had agreed to surrender all claims to the so-called Yazoo Territory, the disputed western area of Georgia out of which Alabama and Mississippi would later be carved.

Instantly legislators at Georgia's state capital at Augusta began celebrating. Now at last the state would be able to sell its western lands and fill its coffers to the brim. Georgia had promised to turn the vast stretch over to the federal government, but now that there was money to be made, the state decided to break its commitment.

Naturally, not all of the money would go to state coffers. Some would have to go to the legislators themselves to persuade them to sell the lands. But that was to be expected.

By November the plans were set. Four companies would be granted the western tract for a price of $500,000, or about one and a half cents per acre. The price was not very high, but the legislators were satisfied. Besides, even if the state did not make a fortune, they would. The four companies, owned by some of the nation's most eminent men, including several congressmen, a state justice, a federal district court judge, and an associate justice of the Supreme Court, had made sure that all legislators who wanted bribes would have them. To Thomas Wylly eight slaves were given; to others patents for Yazoo land. Was Thomas Lanier inclined to refuse a bribe of 50,000 acres? Then give them 75,000 acres. That would ease his conscience.

When the bill selling the lands to the four companies came to a vote early in 1795, it effortlessly sailed through both houses of the legislature. But Governor George Matthews surprisingly vetoed it. Matthews was not averse to selling the western lands—he was as much in favor of speculation and bribery as anyone else—but he didn't believe the lands should be sold at that particular time. If the state waited awhile, he thought, it could easily get a higher price.

But the legislators were not in a waiting mood. They wanted to sell the Yazoo as fast as possible. So after a second round of bribes, they passed another bill. This time the governor signed it.

When the people of Georgia learned that thirty-five million

acres of land had been sold for a pittance and that every legislator but one had accepted a bribe, receiving on average about 50,000 acres of Yazoo land, all hell broke loose. A new legislature was elected, the old law repealed, and at a big bonfire all the documents relating to the fraud were burned. At the state convention of 1798 lawmakers put a clause in the constitution ratifying the legislature's repeal of the Yazoo law.

In 1802, Georgia sold the Yazoo Territory to the federal government for $1,250,000. But the Yazoo controversy was hardly finished. Yazoo stockholders were determined to prove that Georgia had no right to rescind the Yazoo law. After many years of legal battles, the Supreme Court agreed with them. Writing the majority opinion for the case *Fletcher* vs. *Peck,* Chief Justice Marshall ruled that a grant of land was a binding contract which could not be broken regardless of the circumstances under which it was negotiated. In 1814, four years after the Court's decision, the Congress finally awarded the claimants $4,282,151.12.

SOURCE: Nathan Miller, *The Founding Finaglers* (New York: David McKay, 1976), pp. 116–34.

★ ALEXANDER HAMILTON'S AFFAIR ★

In the summer of 1797, Alexander Hamilton responded to charges that he had been personally friendly to a notorious speculator while secretary of the treasury by confessing that he had had an affair with the man's wife. He explained the matter completely in one of the most bizarre statements ever made by an American politician:

"The charge against me is a connection with one James Reynolds for purposes of improper pecuniary speculation. My real crime is an amorous connection with his wife, for a considerable time with his privy and connivance, if not originally brought on by a combination between the husband and wife with the design to extort money from me.

"This confession is not made without a blush. I cannot be the apologist of any vice because the ardour of passion may have made it mine. I can never cease to condemn myself for the pang, which it

may inflict in a bosom [his wife's] eminently intitled to all my gratitude, fidelity and love. But that bosom will approve, that even at
so great an expence, I should effectually wipe away a more serious
stain from a name, which it cherishes with no less elevation than
tenderness. The public too will I trust excuse the confession. The
necessity of it to my defense against a more heinous charge could
alone have extorted from me so painful an indecorum."

SOURCE: Harold Syrett, ed., *The Papers of Alexander Hamilton* (New York:
Columbia University Press, 1974), XXI, 243–44.

★ JUST CAUSE FOR ARREST ★

After Congress adjourned in July 1798, President John Adams traveled home from Philadelphia (then the capital) to Quincy, Massachusetts. When the President and his wife passed through Newark,
New Jersey, the town celebrated the occasion as a holiday. The
church bells rang, people sang, a sixteen-gun salute announced the
President's arrival, and the Association of Young Men manned and
fired an old artillery piece to honor the President's passing.

But to Luther Baldwin, an inebriated old Republican, the presidential presence was of no great importance. (Adams was a Federalist.)
As Baldwin passed John Burnet's dram shop, one of the more plain-
spoken customers who knew Baldwin and his political persuasion
observed, "There goes the President and they are firing at his ass."

"I don't care if they fire through his ass!" Baldwin replied,
with the conviction of a true Republican.

He said too much. Baldwin was immediately arrested under the
new Alien and Sedition laws for muttering an un-American statement. Tried before a circuit court presided over by George Washington's nephew, Bushrod Washington, Baldwin and the other
patron of Burnet's tavern were found guilty of speaking "seditious
words tending to defame the President and Government of the
United States." Both were fined, assessed court costs, and committed to a federal jail until their fines and fees had been paid.

The Federalist-dominated Supreme Court never declared
unconstitutional the Sedition Act, which expired after a few

years. But in 1964, too late to help Baldwin, unfortunately, the Court declared the law unconstitutional in an informal decision.

SOURCE: James Morton Smith, *Freedom's Fetters* (Ithaca, N.Y.: Cornell University Press, 1956), p. 270.

★ JOHN MARSHALL LIKED TO DRINK ★

"Some talk got out about the Justices of the Supreme Court drinking too much. They all lived at the same house in Washington. They did not bring their wives to Washington with them, as the accommodations were frightful. They boarded together at 2½ Street, called Marshall Place. That house still stands. They lived together like a sort of family and discussed their cases all the time; but they had every Saturday as 'consultation day' at the capital.

"There came to be a little talk about the Justices drinking too much, even then. So Marshall said . . . , 'Now, gentlemen, I think that with your consent I will make it a rule of this Court that hereafter we will not drink anything on consultation day—that is, except when it rains.'

"The next consultation day—I think the Court went on the water wagon during the week—when they assembled, Marshall said to [Joseph] Story, 'Will you please step to a window and look out and examine this case and see if there is any sign of rain.' Story looked out the window, but there was not a sign of rain. . . . He came back and seriously said to the Chief Justice, who was waiting for the result, 'Mr. Chief Justice, I have very carefully examined this case, I have to give it as my opinion that there is not the slightest sign of rain.' Marshall said, 'Justice Story, I think that is the shallowest and most illogical opinion I have ever heard you deliver; you forget that our jurisdiction is as broad as this Republic, and by the laws of nature, it must be raining some place in our jurisdiction. Waiter, bring on the rum.'"

SOURCE: Albert Beveridge, "Maryland, Marshall, and the Constitution," in *Proceedings of the Maryland State Bar Association for 1920*, p. 174. Reprinted by permission of the Maryland State Bar Association.

★ JOHN MARSHALL'S SOPHISTRY ★

Thomas Jefferson had this to say about Chief Justice John Marshall: "When conversing with Marshall I never admit anything. So sure as you admit any position to be good, no matter how remote from the conclusion he seeks to establish, you are gone. So great is his sophistry, you must never give him an affirmative answer, or you will be forced to grant his conclusion. Why, if he were to ask me whether it was daylight or not, I'd reply, 'Sir, I don't know. I can't tell.'"

SOURCE: Leonard Baker, *John Marshall* (New York: Macmillan, 1974), pp. 153–54.

★ JEFFERSON'S CONCUBINE ★

Not much was ever heard in scholarly circles about Thomas Jefferson's affair with a slave named Sally Hemings until Fawn Brodie published an account of it in the early 1970s. But the story had been whispered about since Jefferson was president. One of his contemporaries even put the story to verse:

> Of all the damsels on the green
> On mountain, or in valley,
> A lass so luscious ne'er was seen
> As Monticellian Sally.

> **Chorus:** Yankee Doodle, who's the noodle?
> What wife was half so handy?
> To breed a flock of slaves for stock
> A black amour's the dandy . . .

> When press'd by load of state affairs,
> I seek to sport and dally,
> The sweetest solace of my cares
> Is in the lap of Sally.

Chorus: *Yankee Doodle, etc.*

What though she by her glands secretes?
Must I stand, Shill-I-shall-I?
Tuck'd up between a pair of sheets
There's no perfume like Sally.

SOURCE: Hope Ridings Miller, *Scandals in the Highest Office* (New York: Random House, 1973), pp. 72–73. Poem reprinted by permission of Random House, Inc.

★ JEFFERSON'S DEATH REPORTED ★

Thomas Jefferson died in 1826, but in July of 1800 newspapers across the country carried the report that he had died on his estate in Virginia after a brief illness. The report appeared first in Baltimore on June 30 and then made its way up the coast to Philadelphia, New York, and Boston. By the end of the first week of July virtually all Americans had been notified that Jefferson was dead.

Not everyone believed the report. Republicans charged that the news of Jefferson's death was a Federalist trick. With other Americans, they questioned the credibility of the gentleman who had informed the Baltimore paper of the death. The Federalists, on the other hand, secretly hoped that every word of the story was true. With Jefferson out of the picture they might be able to maintain control of the government.

For more than a week the question of Jefferson's health remained in dispute. Finally, the matter was settled when people learned that it was not Thomas Jefferson, Founding Father and vice president, who had died, but one of his old slaves with the same name. Ironically, when Jefferson did die, twenty-six years later, it was in the first week of July.

SOURCE: Charles Warren, *Odd Byways in American History* (Cambridge, Mass.: Harvard University Press, 1942), pp. 127–35.

★ JEFFERSON AND THE PEOPLE'S WALK ★

The story of how Thomas Jefferson walked to his own inauguration in 1801 wearing a gray homespun suit has been told by historians a thousand times. And the story is true. But the very truth of the story has contributed to an untrue impression of the third president. The Virginian's example of simplicity was completely unintended. Jefferson walked to the inauguration and wore plain clothes only because bad weather had delayed the arrival of a new $6,000 carriage and an expensive velvet suit.

SOURCE: Edna Colman, *Seventy-five Years of White House Gossip* (Garden City, N.Y.: Doubleday, Page, 1925), p. 76.

★ THOMAS JEFFERSON'S LIST OF DIFFERENCES ★
BETWEEN THE NORTH AND THE SOUTH

Northerners	Southerners
cold	fiery
sober	voluptuary
laborious	indolent
independent	unsteady
jealous of their own liberties, and just to those of others	zealous for their own liberties, but trampling on those of others
interested	generous
chicaning	candid
superstitious and hypocritical in their religion	without attachment or pretensions to any religion but that of the heart

SOURCE: Mary Cable, *American Manners and Morals* (New York: American Heritage Publishing Company, 1969), p. 95.

★ OTHER DISCOVERIES OF LEWIS AND CLARK ★

According to their travel journals, Lewis and Clark found more than merely the land of the Louisiana Purchase on their explorations of 1804–1806. Venereal disease was a major problem throughout the expedition. Because of the different sexual customs practiced by some Indian tribes, the explorers did not remain totally abstinent during their three-year trip. Meriwether Lewis, leader of the expedition, hinted at the situation when he wrote that a Shoshone warrior "will for a trifle barter the companion of his bed for a night or longer if he conceives the reward adequate." Lewis added, however, that the Shoshones were "not so importunate that we should caress their women as the sioux were."

In another journal entry, dated March 15, 1806, Lewis recorded that the expedition was "visited this afternoon by Delaskshelwilt a Chinnook Chief his wife and six women of his nation . . . this was the same party that had communicated the venerial to so many of our party in November." For this particular visit, Lewis gave strict orders that his men were not to behave in any manner that might infect or reinfect them with VD. But for the most part Lewis and Clark did not even attempt to keep their men and the Indian women apart. "To prevent this mutual exchange of good officies altogether," wrote Lewis, "I know it impossible to effect, particularly on the part of our young men whom some months abstinance have made very polite to those tawney damsels."

It is unlikely that Lewis and Clark themselves totally abstained from relations with Indian women. In the middle of January 1805, William Clark wrote that a leading Minnetaree war chief had visited Fort Mandan, the expedition's winter headquarters, with his squaw and "requested that she might be used for the night." Clark added that the chief's "wife [was] handsome," but otherwise maintained a discreet silence on the topic.

Mercury was the cure for VD in the early 1800s, and it was reported that the expedition spent many an hour rubbing the medication directly into the skin or swallowing a mercury derivative in pill form. Either treatment was guaranteed to prove the truthfulness of the popular adage: "A night with Venus, a lifetime with Mercury."

SOURCES: Paul R. Cutright, *Lewis and Clark* (Urbana, Ill.: University of Illinois Press, 1969), p. 254; Reuben Gold Thwaites, ed., *Original Journals of the Lewis and Clark Expedition* (1904–5; rpt. New York: Arno, 1969), IV, 170; John Bakeless, *Lewis and Clark* (New York: William Morrow, 1947), pp. 182–83.

★ JUST PUNISHMENT OF A WHITE MAN? ★

On Sunday, October 14, 1804, John Newman, a member of the Lewis and Clark Expedition, was severely whipped for insubordination. His punishment was witnessed by the chief of the Arikara Indians, who expressed alarm at what he saw. The chief was so affected by the scene that he literally broke down and cried. Clark explained that Newman had been insubordinate and had to be punished. The chief agreed that insubordination could not be tolerated. He himself had punished his own men for being insubordinate. But he did not believe in whippings. His people never even whipped their own children. The chief declared that if Newman had been one of his men, the soldier would not have been whipped—he would have been killed. Death was the only sure cure for insubordination.

SOURCE: Bernard DeVoto, ed., *The Journals of Lewis and Clark* (Boston: Houghton Mifflin, 1953), p. 51.

★ YORK AND THE INDIANS ★

For the Indians the most unusual item carried by Lewis and Clark was not the few trinkets the explorers handed out, or their guns, but one of the members of the expedition itself. York, William Clark's manservant, was a black man. Early trappers and traders, even in the most remote areas, had made the Indians aware of the white man. But a man with black skin was a truly startling phenomenon. The reaction of Le Borgne, the grand chief of the Minnetarees, typified Indian surprise and disbelief. Meriwether Lewis recorded that on Le Borgne's first visit to the expedition's winter headquarters at Fort Mandan (now Bismarck, North Dakota) on March 9, 1805, "the chief observed that some foolish young men of his nation had told him there was a person among us who was quite black, and he wished to know if it could be true." When York appeared, Chief Le Borgne was dumbfounded. After examining the Negro closely, Lewis recorded, the Minnetaree chief "spit on his [own] finger and rubbed [York's] skin in order to wash off the paint." Not until York uncovered his head and displayed a short crop of curly hair could the dubious Le Borgne be persuaded that the black man was not a painted white man.

Indian admiration for this strange visitor was genuine. In discussing sexual encounters with Indian women, Lewis recorded that "the black man York participated largely in these favours, for instead of inspiring any prejudice, his color seemed to procure him additional advantages from the Indians, who desired to preserve among them some memorial of this wonderful stranger." Lewis reported how one Ricara Indian invited York to his house, presented the Negro to his wife, and then retired outside by the door. When one of York's companions came by and inquired about the Negro, the Indian would not let the man enter the house until a suitable time had elapsed. Reportedly, York sired a number of mixed-blood offspring throughout the three-year trip.

SOURCE: Meriwether Lewis, *History of the Expedition of Captains Lewis and Clark* (Chicago: A. C. McClurg and Company, 1924), I, 180.

★ HOME REMEDIES THAT KILLED ★

In the early nineteenth century the odds were against the person who recklessly chose to receive medical treatment when sick. Doctors were not licensed then, and a good many of them were quacks. But even the respected ones did not really know what they were doing. The doctors who were most to be feared, as a matter of fact, were those who had worked hardest to develop cures. In general, the best doctors were those who resorted simply to common sense. The patient who took George Washington's advice to let a cold "go as it came" was probably just as well off or better as the person who called for a doctor.

Exceedingly dangerous to patients were the widely touted home remedies. Some were harmless, of course, but many were absolutely lethal. The following recipe for killing worms in children carried its own warning of risk: "Take sage, boil it with milk to a good tea, turn it to whey with alum or vinegar, and give the whey to the child, if the worms are not knotted in the stomach, and it will be a sure cure. If the worms are knotted in the stomach, it will kill the child."

SOURCE: Mary Cable, *American Manners and Morals* (New York: American Heritage Publishing Company, 1969), p. 148.

★ A DIFFERENT DAY ON THE ROAD ★

Toll roads have been a part of the American landscape since the early part of the nineteenth century. Toll traffic, however, has changed. The turnpike authority for the sixty-eight-mile road from Schenectady to Utica, New York, posted the following charges in 1809:

Sheeps per score	8¢
Hogs per score	8¢
Cattle per score	18¢
Horses per score	18¢
Mules per score	18¢
Horse and rider	5¢
Tied horses, each	5¢
Sulkies	12½¢
Chairs	12½¢
Chariots	25¢
Coaches	25¢

Phaetons	25¢
Two-horse stages	12½¢
Four-horse stages	18½¢
One-horse wagons	9¢
Two-horse wagons	12½¢
Three-horse wagons	15½¢
Four-horse wagons	75¢
Five-horse wagons (wheels under six inches)	87½¢
Six-horse wagons (wheels under six inches)	$1.00
One-horse cart	6¢
Two-ox cart	6¢
Three-ox cart	8¢
Four-ox cart	10¢
Six-ox cart	14¢
One-horse sleigh	6¢
Two-horse or two-ox sleigh	6¢
Three-horse or three-ox sleigh	8¢
Four-horse or four-ox sleigh	10¢
Five-horse or five-ox sleigh	12¢
Six-horse or six-ox sleigh	14¢

SOURCE: Alice Morse Earl, *Stage-Coach and Tavern Days* (New York: Macmillan, 1900), pp. 237–38.

★ THE RED BARN ★

Why are barns painted red? In the early nineteenth century farmers learned that the color red absorbed sunlight extremely well and was useful in keeping barns warm during winter. The farmers made their red paint from skim milk mixed with the rust shavings of metal fences and nails.

Source: Wilson Clark, *Energy for Survival* (Garden City, N.Y.: Anchor, 1975), p. 576.

★ Vermont Slides Past the Law ★

As the Napoleonic Wars in Europe escalated, British and French harassment of neutral American shipping greatly disturbed American leaders. Finally, after an English frigate fired on the USS *Chesapeake* in December 1807, President Thomas Jefferson invoked the Embargo Act, which outlawed trade with warring countries.

In Vermont news of the embargo was received apathetically. Since there were no ocean ports in the new state, the law hardly seemed worth getting upset about. But when the Green Mountain State learned that their brisk trade with Canada was also forbidden by the embargo, they set about creating ways to overcome the unpopular law.

Some smugglers built docks on Lake Champlain along the U.S.–Canadian border. American ships would then dock and unload on the south side while British vessels reloaded the cargo from the north side, just across the border and out of reach of U.S. customs. Other smugglers were more brazen. In late June 1808, a group of them stole the U.S. revenue cutter that regularly patrolled the lake.

A few Vermonters carried goods to buildings erected on hilltops directly across the Canadian border. After the house was "loaded up," a specific stone or piece of wood was removed and the building would "accidentally" slide down the hill into Canada.

Extremely unpopular, the Embargo Act was repealed in 1809.

Source: H. N. Muller, "Smuggling into Canada," *Vermont History*, XXXVIII (Winter 1970), passim.

★ British Democracy in 1812 ★

Washington, D.C., has long been known as the capital of democracy. Even the British acknowledged this fact when they occupied the city during the War of 1812. Admiral Cockburn, the leader of the British expeditionary force in the city, took the Speaker's chair in the House of Representatives and polled the assembled officers and men. "Gentlemen," he said, "the question is, Shall this harbor of Yankee Democracy be burned? All in favor of burning it will say aye!"

The vote was unanimous. "Light up," ordered Cockburn, and the capital was burned.

SOURCE: Glenn Tucker, *Poltroons and Patriots* (Indianapolis: Bobbs-Merrill, 1954), p. 556.

Present-day picture of a column outside the old Supreme Court chamber showing bullet holes supposedly made by the British during their attack on the capital in 1814. (U.S. Senate Historical Office.)

★ JOHN RANDOLPH'S ADVICE TO A NEIGHBOR ★

John Randolph, Virginia congressman and senator, was one of the most outrageous men of his time. He used to bring his dogs onto the floor of Congress, and on several occasions engaged in brawls with other representatives right in the Capitol. Stories about his vituperation and wit are countless:

"During one of the last years of his life, Mr. R. was an attendant on the sessions of the Virginia Legislature, when a bashful, back-country planter met the eccentric orator in the lobby and endeavored to introduce himself. 'Mr. Randolph,' said he, fumbling and scraping with especial awkwardness, 'I live only fifteen or twenty miles from you—I pass your plantation quite often.'—'Sir,' said John, regarding him from head to foot with infinite scorn, 'you are welcome to *pass* it as often as you please.'"

SOURCE: Horace Greeley, ed., *The Tribune Almanac for the Years 1838 to 1868* (New York: New York Tribune, 1868), I, 42.

★ RANDOLPH CHASTISES A COLLEAGUE ★

During the debate over the Missouri Compromise of 1820, Randolph stood up repeatedly to oppose the measure. Almost every time he began speaking, however, he was interrupted by Philomen Beecher of Ohio, who would move the "previous question." The Speaker of the House would call Beecher to order, and Randolph would proceed with his speech. But finally, after yet another interruption from Beecher, the Virginian was moved to remark: "Mr. Speaker, in the Netherlands, a man of small capacity, with bits of wood and leather, will, in a few moments, construct a toy that, with the pressure of the finger and thumb, will cry 'Cuckoo! Cuckoo!' With less ingenuity, and with inferior materials, the people of Ohio have made a toy that will, without much pressure, cry 'Previous question, Mr. Speaker!'" Beecher turned red and John Randolph was never again bothered by the cry "Previous question, Mr. Speaker!"

SOURCE: Leon A. Harris, *The Fine Art of Political Wit* (New York: Dutton, 1964), pp. 57–58.

★ POTENT WORDS ★

John Randolph of Virginia and Tristam Burges of Rhode Island were not ones to confine their remarks on the Senate floor to politics. Burges once alluded to Randolph's sexual impotency, saying, "But I rejoice that the Father of Lies can never become the Father of Liars." "You boast of a quality," rejoined Randolph, "in which all slaves are your equal and every jackass your superior."

SOURCE: Edward Boykin, ed., *The Wit and Wisdom of Congress* (New York: Funk & Wagnalls, 1961), p. 161.

★ NEVER TURN OUT FOR SCOUNDRELS ★

"One day [Henry] Clay met his disagreeable enemy, [John] Randolph, on the sidewalk. The cranky old Virginian came proudly up, and occupying most of the sidewalk hissed: 'I never turn out for scoundrels!' 'I always do,' said Clay, stepping aside with mock politeness."

SOURCE: Melville D. Landon, *Eli Perkins: Thirty Years of Wit* (New York: Cassell, 1891), p. 142.

★ PRINCE AMONG SLAVES ★

Born in 1762 in the forbidden city of Timbuktu, Ibrahima was the son of a powerful African monarch. A devout Moslem, he was educated in Timbuktu and elsewhere. His father was king of the Futa Jalon area in the present-day Republic of Guinea. Ibrahima had a wife and family.

In 1781 members of the Futa tribe found an English surgeon, John Coates Cox, hopelessly wandering in the African bush. Cox had left his ship to hunt, but had become hopelessly lost. He was sick and starving. For over six months Cox stayed in Futa Jalon, until he fully regained his health. He owed his life to the hospitality of Ibrahima's father.

Seven years later Ibrahima, a colonel in his father's army, was defeated in war and sold into slavery. Eventually, he became the slave of Thomas Foster, a small farmer in Natchez, Mississippi.

In Mississippi, Ibrahima was called "Prince," because of his rumored royal heritage. Once, a few years after he arrived in America, another slave, who presumably had been a member of the Futa tribe of Africa, recognized him on the streets of Natchez and instinctively dropped his head to the ground, crying, "Abduhl Rahahman!"

One day in 1807, while selling potatoes in Washington, Mississippi, a small village near Natchez, Prince saw John Cox riding down the street on a horse. Cox had immigrated from England to New York in 1786, then moved to North Carolina, and later to Mississippi. Over twenty-five years had passed since the Futa had saved Cox's life. When Cox recognized Ibrahima, he leaped off his horse and embraced the African.

Ibrahima was an extremely valuable worker, however, and Thomas Foster repeatedly refused Cox's offer to buy the slave. For the next nine years, until Cox's death, the surgeon and Ibrahima saw each other often, but Ibrahima remained in slavery. Spurred on by Cox's numerous stories about Africa, however, Prince's reputation grew and he became something of a legend in the Natchez area.

In 1826, after years of persistent urgings by Natchez's local printer, Colonel Marshalk, Prince wrote a verse of the Koran in Arabic and Marshalk mailed it to Morocco. For some reason, Marshalk believed Prince was originally from the North African kingdom, and thought it would be exciting to find out if the old slave could make contact with his homeland. Half a year later, Abdal-Rahman II, king of Morocco, read Prince's quote and offered to pay for the Moslem's freedom and traveling expenses to Morocco. When Secretary of State Henry Clay sent the Moroccan king's request to Mississippi, Thomas Foster emancipated the old slave and his wife for free.

Once the manumitted pair reached Washington, Prince tactfully announced that he was not from Morocco, and arrangements were made to send him instead to the American Colonization Society's colony of Liberia, on the central African coast. Before he left America, Prince traveled to Philadelphia, New York, and Boston and held large rallies, hoping to raise enough money to purchase the freedom of his nine children and grandchildren, who remained in slavery in Mississippi. With the proceeds from these engagements, Prince's son, his wife, and their five children were able to emigrate to Liberia in 1830. Another son, Lee, was later able to move to the colony. Prince's other offspring remained in slavery.

On May 15, 1828, Prince, the son of an African king and a slave for forty years, met personally with President John Quincy Adams in Washington. Later that year Prince returned to Africa.

Leadership of Futa Jalon, after forty years, still belonged to Ibrahima's family. When they received word that their lost relative was in Liberia, a caravan was dispatched immediately to carry him home. Unfortunately, on July 6, 1829, after a long illness, Ibrahima died. The Futa caravan, which carried six to seven thousand dollars in gold dust for the manumission of Ibrahima's family in Mississippi, was only one hundred and fifty miles away. When news of Ibrahima's death reached the caravan, it turned around and traveled back to Futa Jalon.

SOURCE: Terry Alford, *Prince among Slaves* (New York: Harcourt Brace Jovanovich, 1977), *passim*.

★ JOHN QUINCY ADAMS'S TENEMENT ★

When John Quincy Adams was an old man, he responded to questions about his health with humor. To an old friend he remarked, "I inhabit a weak, frail, decayed tenement; battered by the winds and broken in upon by the storms, and, from all I can learn, the landlord does not intend to repair."

OLD HICKORY TO OLD ROUGH-AND-READY

★ ≋

"This country is filling up with thousands and millions of voters, and you must educate them to keep them from our throats."

—RALPH WALDO EMERSON

★ SCRAPBOOK OF THE TIMES ★

- James Fenimore Cooper once remarked, "'They say' [is] the monarch of this country."
- Henry Clay delivered an hour-long speech in the Senate on his dead bull, Orozimbo.
- When John Quincy Adams lost to Andrew Jackson in the election of 1828, the town of Adams, New Hampshire, changed its name to Jackson. The town had been named in 1800 to honor the election of John Adams.
- Until the 1830s, Americans did not eat tomatoes. Up to that time tomatoes were believed to be poisonous and were used only as decorations. They were known as "love apples."
- During his time as a congressman Davy Crockett made repeated attempts to abolish West Point, which he believed was a haven for the sons of aristocrats.
- In 1832, Henry Clay presented to the Senate a petition requesting land for two Kentuckians who claimed to have discovered "the secret of living forever." The petition was denied.
- Women were not allowed on the floor of the Senate until the 1830s.
- Although he was a fine orator and brilliant senator, Daniel Webster was, to say the least, a careless spender. Often he had to take money from businessmen to keep his books balanced. Occasionally this led to a conflict of interests, as when Web-

ster defended the Bank of the United States on the floor of the Senate at the same time that he was receiving money from the head of the Bank, Nicholas Biddle.

- The first person to steal a million dollars from the federal government was Samuel Swartwout, collector of the Port of New York, who, after embezzling the money, fled to Europe and safety.

- In January 1835 the United States became the only major nation in modern history to pay off completely its national debt. This was accomplished through the sale of public lands in the West. Eighteen thirty-five was the only year America has had no debt.

- While Martin Van Buren was vice president, he presided over the Senate wearing a pair of pistols, as a precaution against the frequent outbursts of violence.

- Washington Irving coined the phrase "The Almighty Dollar" in 1837 in his book *The Creole Village*.

- William Henry Harrison delivered the longest inaugural address of any president and served the shortest time: one month, to the day. In his address he made the prophetic remark that he would not be a candidate for a second term.

- John Tyler was on his knees playing marbles when informed that he had become president on the death of Harrison.

- Tyler was the first president against whom a resolution of impeachment was drawn. The resolution, which was defeated by a vote of 127 to 83, came about after a congressional committee chaired by ex-president John Quincy Adams accused Tyler of abusing the power of the veto.

- The word "millionaire" was coined in 1843 by a newspaper reporter in an obituary of Pierre Lorillard, banker, landlord, and tobacconist.

- Baseball was invented in 1845 by a genteel group of New York businessmen, who once snobbishly refused to play the game with a team of "greasy mechanics" from Brooklyn.

- When Sarah Childress Polk became first lady, she immediately banned dancing from the White House. For four years no one danced one step there.

- The Smithsonian Institution was founded in 1846 and named in honor of an Englishman, James Smithson.
- A week before he died, James Polk fulfilled a lifelong promise to his wife and was baptized.
- Beginning in the 1840s, Americans built thousands of miles of roads out of wood planks. The planks were laid side by side to provide a mud-free surface for carriages and wagons. Over a fifteen-year period more than 7,000 miles of plank roads were constructed.
- John Jacob Astor once remarked, "A man who has a million dollars is as well off as if he were rich."
- U. S. Grant, remarking in 1846 on the short-range muskets used during the Mexican War: "A man might fire at you all day without your finding it out."

★ BIRTH CONTROL IN EARLY AMERICA ★

In 1823 new methods of birth control began to be popularized. Although mentioning the traditional *coitus interruptus* as a useful procedure, handbills first printed in London in that year urged couples to "do as other people do" and subscribe to a new method in which "a piece of soft sponge is tied by a bobbin or penny ribbon, and inserted just before the sexual intercourse takes place, and is with drawn again as soon as it has taken place." The instructions went on to say that "many tie a piece of sponge to each end of the ribbon, and they take care not to use the same sponge again until it has been washed." The uncirculated draft of the handbill mentioned as an alternative to the sponge a tampon of "lint, fine wool, cotton, flax, or what may be at hand." The sponge method gained some acceptance in the United States.

Much more popular in America, however, were the ideas of Dr. Charles Knowlton, one of the initiators of the birth-control movement in this country. In his popular publications, which first appeared in 1832 and ran through many editions, he recommended douching as the chief method of control. Solutions of either alum or "astringent vegetables [such] as white oak bark, hemlock bark, red rose leaves, green tea, and raspberry leaves or roots" were suggested. Under special circumstances Knowlton prescribed special douches. When the membranes of the vagina were inflamed, a solution of sulfate of zinc was advised. When "relaxation was present," Knowlton thought a combination of zinc and alum was most suitable. Or, if there was "tenderness of the parts," Knowlton recommended a "solution of sugar of lead."

Of course, homes in 1832 were not well heated. But this was not a problem for Knowlton's birth-control formula. In mixing the solutions, the doctor advised that a little "spirits" be added to prevent freezing.

SOURCE: Norman E. Himes, *Medical History of Contraception* (New York: Gamut, 1963), pp. 216–17, 227–28.

★ FAMOUS AMERICAN DUELS ★

Following is a partial list of leading Americans who at one time or another made or accepted the challenge to a duel:

Benedict Arnold (1792)

Thomas Hart Benton (1813, 1817)

John C. Breckinridge (1854)

Aaron Burr (1804)

Henry Clay (1808, 1826)

De Witt Clinton (1802, 1803)

William H. Crawford (1802)

Stephen Decatur (1801, 1803, 1820)

Nathanael Greene (1785)

Alexander Hamilton (1797, 1804)

Samuel Houston (1845)

Andrew Jackson (involved in more than 100 duels)

John Jay (1785)

Charles Lee (1778)

Abraham Lincoln (1842)

James Madison (1797)

John Randolph (1807, 1826)

Winfield Scott (1817)

James Wilkinson (1807)

SOURCE: Lorenzo Sabine, *Notes on Duels and Duelling* (Boston: Crosby, Nichols & Company, 1856), *passim*.

★ ANDREW JACKSON ARMS HIS SLAVES TO KEEP ★ THEM IN SLAVERY

Andrew Jackson was a slaveowner. He was also one of the brashest men ever to become president.

In 1811, four years before his victory over the British at the Battle of New Orleans, Jackson was returning to Nashville from Natchez, Mississippi, with twenty-six slaves. The road, the famous Natchez Trace highway, crossed directly through Choctaw territory. An agent of the acting governor of the territory, in response to a great number of complaints about whites helping runaways escape, had begun requiring certificates of ownership for every slave taken through the Indian area. Jackson believed the agent was harassing whites by this policy. The future president even asserted that the agent's motive was to take possession of uncertified slaves so that he could put them to work on his own plantation. Old Hickory, of course, had no intention of being a victim of such "despotism."

On entering the Choctaw land, Jackson was stopped and asked for proof that the slaves he was traveling with were not runaways. Jackson knew who owned his slaves and saw no need to prove his ownership to anyone. He immediately recognized the request as a pretext for illegally taking possession of his slaves.

As a citizen of the United States, he coarsely replied, the only proof he needed was an honest face and a good reputation!

This response did not sit well with the agent, who armed his twenty men to stop Jackson from crossing the Indian territory. But the ever-resourceful Jackson was not easily defeated.

Traveling with only one other white man, Jackson returned to his twenty-six slaves, removed their chains, passed out axes and clubs, and then marched his newly formed "platoon" past the bemused agent. No one dared interfere with the two half-mad whites and the army of twenty-six unshackled, club-wielding slaves. Once safely into Tennessee, Jackson collected the weapons, rechained his slaves, and continued his journey home. No incident occurred with the blacks. Later, Jackson sold the majority of the twenty-six slaves.

Back in Nashville, Jackson fumed over his treatment at the hands of the agent. Immediately, he fired off letters to Tennessee congressmen and senators and other officials, vociferously denouncing the situation where "citizens are to be threatened with chains and confinement for peaceably travelling a road." Jackson demanded that action be taken against the agent to make sure that never again would a white man's "right" to his slaves be questioned. Needless to say, Old Hickory's letters reflect no sense of irony over the fact that the very men who made possible the future president's heroics were themselves condemned to a life of "chains and confinement."

SOURCES: Andrew Jackson, letter to Willie Blount, January 25, 1812, National Archives, Washington, D.C.; James Parton, *Life of Andrew Jackson* (New York: Mason, 1859), I, 349–60.

★ OLD-FASHIONED AMERICAN MANNERS ★

When Andrew Jackson in late 1832 was on his way to his inauguration as president, he stopped in Cincinnati. There he was mobbed by well-wishers, each eager to press the hand of the great general. The outpouring warmed Jackson's heart and helped him forget that his beloved wife, Rachel, had died just a few weeks before. But as he was leaving, someone in the crowd painfully reminded him of her death.

"General Jackson, I guess?" a greasy fellow asked him. "Why, they told me you was dead."

"No," Jackson replied, "Providence has hitherto preserved my life."

"And is your wife alive too?" the man asked. Jackson sadly responded that she wasn't.

"Aye," said the man, with a look of complete self-satisfaction, "I thought it was the one or the t'other of ye."

SOURCE: Frances Trollope, *Domestic Manners of the Americans*, ed. Donald Smalley (1832; rpt. New York: Knopf, 1949), p. 145.

★ HIS CHANCE OF LIVING WAS SMALL ★

The first attempt to assassinate a president occurred on January 30, 1835, when Andrew Jackson was shot at by Richard Lawrence, a deranged Englishman. Jackson was at the House of Representatives attending the funeral of Congressman Warren Davis of South Carolina when Lawrence quietly walked up to him and from six feet away fired two pistols. Remarkably, Jackson wasn't hurt. Only the caps of the guns exploded, though both had been correctly loaded. When one of the guns was recapped later, it discharged perfectly. An expert on guns estimated that the chance of the pistols not firing was one in a hundred and twenty-five thousand. Lawrence, who claimed he was an heir to the British throne, was immediately apprehended, convicted, and committed to an insane asylum.

SOURCE: Marquis James, *The Life of Andrew Jackson* (Indianapolis: Bobbs-Merrill, 1938), II, 390–91.

★ BENTON WRITES A BIG BOOK ★

"The animosity between [Senator Henry S.] Foote, of Mississippi, and [Senator Thomas Hart] Benton, of Missouri, was well known.

It is a matter of record. . . . Mr. Foote said that he would write a little book in which Mr. Benton would figure very largely. Mr. Benton heard of this and replied in his characteristic way to his informant: 'Tell Foote that I will write a very large book in which he shall not figure at all.'"

SOURCE: *Harper's Magazine*, September 1859, p. 569.

★ LET THE ASSASSIN SHOOT ★

"One morning shortly after Benton had forbidden Foote ever to mention his name again [on the floor of Congress] the latter arose in his place and began a severe attack on Benton. As soon as his name was mentioned Benton started for him in a menacing manner, whereupon Foote hopped out into the big aisle, drew a revolver, cocked it, and leveled it at Benton. A couple of Senators, seeing what was about to happen, grabbed Benton and tried to stop him, which they could not do on account of his great strength. They could only retard his progress. He kept on. As he approached close to Foote, the latter, with his cocked revolver still in hand, began to retreat down the big aisle till he reached the Vice President's stand, Benton tugging along after him as fast as he could, with two Senators holding on to him. When they came to a standstill Benton tore his shirt open and exclaimed: 'Let the assassin shoot! He knows that I am not armed!'"

Foote did not fire and the crisis passed. But the event was important; it was the only time in American history that one Senator ever drew a pistol on another on the floor of the Senate.

SOURCE: Champ Clark, *My Quarter Century of American Politics* (New York: Harper & Brothers, 1920), II, pp. 255–56. Reprinted by permission of Harper & Row, Publishers, Inc.

★ BENTON BOWS BEFORE GOD ★

"It is said that one day [Thomas Hart Benton] intended to answer a speech of Calhoun's, but hearing that Calhoun was prostrated by illness and could not be present, he announced, 'Benton will not speak to-day, for when God Almighty lays his hands on a man Benton takes his off.'"

SOURCE: William M. Meigs, *The Life of Thomas Hart Benton* (Philadelphia: Lippincott, 1904), p. 452.

★ THE BURNING OF BENTON'S HOUSE ★

A story about Thomas Hart Benton by his daughter:

"When my father's Washington house was burned it gave so much pain to every one that both houses adjourned and the silent, helpless crowd bared their heads to my father as he came to the ruin of his home. 'It makes dying easier,' he said to me; 'there is so much less to leave.'"

SOURCE: Charles Shriner, ed., *Wit, Wisdom, and Foibles of the Great* (New York: Funk & Wagnalls, 1918), p. 35.

★ WHEN PEOPLE BELIEVED IN THE ★ MAN IN THE MOON

In 1835 the New York *Sun* perpetrated the greatest hoax in American history. One hundred and thirty-four years before man stepped on the moon, the *Sun* had Americans believing that there was life on the moon. The newspaper reported that an English astronomer, Sir John Herschel, had published articles in the Edinburgh *Journal of Science* confirming the age-old suspicion that creatures inhabited the moon. The creatures were described in terrifying detail, from their furry bodies to their bat wings.

Other newspapers picked up the story, and for weeks people everywhere in the country believed it was completely true. A group of scientists from Yale University even traveled to New York to inspect the original accounts written by Herschel. Finally, the *Sun* revealed that John Herschel didn't exist and that the *Journal of Science* was fictitious. By that time circulation of the paper had soared to 19,360, largest in the world.

SOURCE: Curtis MacDougall, *Hoaxes* (New York: Macmillan, 1940), pp. 229–30.

★ CHANGING PRICE OF LAND IN CHICAGO ★

The phrase "boom town" must have been invented for Chicago. Between 1830 and 1840 the population of the city grew eightfold, while the price of land skyrocketed. In 1820 an acre cost a mere $1.25. By 1832, $100 lots were common. And in 1834 prime property went for $3,500. Land by the lake cost even more: over $20,000 an acre.

Source: Alex Groner, *The American Heritage History of American Business and Industry* (New York: American Heritage Publishing Company, 1972), p. 93.

★ DAVY CROCKETT AND THE MONKEY ★

"Davy Crockett was once attending a menagerie exhibition in Washington, and dilating to some friends on the similarity of countenance between one of the monkeys and a brother Member of Congress. He looked up, and behold! the Member in question was a quiet listener to his discourse!—'I suppose, Mr. W———,' said Davy, 'that I ought to apologize; but I can't tell whether to you or the monkey!'"

Source: Horace Greeley, ed., *The Tribune Almanac for the Years 1838 to 1868* (New York: New York Tribune, 1868), I, 42.

★ REMEMBERING THE ALAMO ★

The Alamo, the old Spanish mission founded at San Antonio, Texas, in 1718, was the site of one of the most famous sieges in all of American history. But the 1836 stand has been surrounded by legends that obscure the truth of what happened.

The historic old church that stands at the Alamo today looked quite different in 1836. Construction of the Catholic mission was basically completed by 1750. Eighty-six years later the mission was an abandoned, dilapidated building. The ceiling of the main room of the church was almost entirely caved in, and the center of the building was filled with rubble and debris. The heroic defenders of the Alamo used some of the debris to construct a cannon emplacement on the back wall of the church. In 1849 a Major E. J. Babbit purchased the building in the name of the United States. While cleaning out the rubble, he found two decomposed bodies, reportedly victims of the battle thirteen years earlier.

The famous church formed only a small part of the entire Catholic mission. Actually, the Alamo today is only one-ninth the size of the fortifications defended in 1836. Colonel Travis, Jim Bowie, Davy Crockett, and the other men defended the entire mission. If standing today, the old mission would include not only the church but the Alamo Plaza city park, several busy streets, and a good number of the area's souvenir shops and tourist-catering taco stands.

The reputations of some of the Texas defenders were not of the highest caliber. Colonel Travis, contemporary evidence indicates, personally killed a man for making advances on his wife. A short time after the murder, he left his pregnant wife and two-year-old son in Alabama and emigrated to Texas. In the oath he took as a colonist, Travis lied, claiming in one case to be a widower and in another instance to be a bachelor. Travis expected reinforcements throughout the Alamo siege. Otherwise, according to some unfriendly accounts, he and his men would have fled the old mission at the first sight of the advancing Mexicans.

Man for man, there was no comparison between the Texas volunteers and the Mexican army. As one historian put it, all a private in the Mexican army had to look forward to was "bad leadership, poor pay, and no glory." The Mexican draft system was so corrupt only the poorest, most forlorn "citizens" could not bribe their way out of service. A great number of General Santa Anna's troops were Mayan Indians who could not even understand Spanish. Away from the tropics for the first time, many ill-supplied Mayans died of exposure in the cold Texas weather. Santa Anna, the self-proclaimed "Napoleon of the West," issued his troops English guns left over from the days of the Battle of Waterloo! Their range was barely seventy yards, while the Texas long rifles were accurate at two hundred yards. Modern-day military historians rank Santa Anna as one of the "top five worst military commanders of all time." Still, with over 6,000 troops surrounding the 182 Texans, the odds favored Santa Anna.

Perhaps the most famous story about the Alamo was the incident in which Commander William Travis drew a line in the dirt with his sword. Only those willing to die for Texas independence, Travis announced to the garrison, should step across the line and defend the Alamo. Jim Bowie, too sick to move from his cot, called over some of his "boys" and had them carry him across Travis's line.

Despite the legend, the story about Travis and Bowie is at best highly questionable. One of the men at the mission that day was Moses Rose, a veteran of the Napoleonic Wars, who fled the Alamo a day or two before the massacre. Later that year he reportedly stayed with the parents of William P. Zuber, who was at the time off fighting in Sam Houston's army. Before Rose left the household, he told Zuber's parents about Colonel Travis's speech, the line in the dirt, Bowie's reaction, and how he had personally decided to go over the wall instead of the colonel's line. After Sam Houston's victory at San Jacinto, Zuber returned home and was told about Rose's story. Thirty-five years later Zuber wrote a third-hand account from memory and published the piece, entitled "An Escape from the Alamo," in the *Texas Almanac* of 1873. From there the story was widely circulated and soon became accepted by the public as fact.

One man who did not follow Rose over the wall but, contrary to legend, did not go down "fighting like a tiger" was Davy Crockett. Overwhelming contemporary evidence indicates that in the waning moments of the great struggle Crockett, along with several other Tennesseans, was captured and taken before General Santa Anna. Ignoring pleas from Mexican general Castrillon to show mercy, Santa Anna immediately ordered the troops standing nearest him to execute the remaining defenders.

SOURCE: Dan E. Kilgore, *How Did Davy Die?* (College Station: Texas A. & M. Press, 1978), *passim;* Walter Lord, *A Time to Stand* (New York: Harper & Brothers, 1961), pp. 40, 60; William Zuber, *My Eighty Years in Texas* (Austin: University of Texas Press, 1971), p. xi.

The abandoned Alamo church as it looked in 1846. (Executive Document 32, U.S. Senate, 31st Cong., 1st sess.)

★ The Yellow Rose of Texas? ★

The song "The Yellow Rose of Texas" honors "the sweetest rose-bud that Texas ever knew." The "sweetest little rose" the song refers to, however, was not a typical Texas belle but a young slave named Emily.

During the 1836 Texas Revolution, as Santa Anna's ragged army approached San Antonio and the Alamo, the general, who had a great desire for women, learned of a beautiful girl who lived in the area. But the girl's mother was a good Catholic, whose husband had been a former officer in the Mexican army. If General Santa Anna wanted her daughter, she insisted, he would have to marry her. Santa Anna's moral convictions were more flexible. But if the old lady wanted a wedding for her daughter, a wedding—of sorts—she would get. Santa Anna had one of his soldiers borrow the necessary Catholic vestments and missal and ordered him to dress up as a priest. Then Santa Anna was "married" by the "priest" in the general's own headquarters. The "honeymoon" lasted throughout the army's stay in San Antonio and ended sixty miles east at Gonzales, Texas, where the royal carriage could not ford the swollen Guadalupe River. The "bride" and carriage were then sent back to San Luis Potosi, Mexico. Santa Anna, of course, traveled on with his army in search of Sam Houston and the rebellious Texans.

But the general was not one to be long without a companion. At Morgan's Point, Texas, Santa Anna took the slave Emily from the household of Colonel James Morgan and made her his "serving maid." She served him too well. On April 21, 1836, Santa Anna and Emily were partying in the general's gaudy, carpeted marquee at San Jacinto, Texas. Champagne cases were stacked outside the entrance. Suddenly, battle cries filled the air as Sam Houston's Texans surprised the Mexicans. The battle was over practically before it could begin. Santa Anna had been too involved with Emily to organize or rally resistance. He barely escaped the battlefield, clad only in his drawers and red slippers. In twenty minutes, 800 Texans had captured or killed the entire 1,500-man Mexican army. The next day General Santa Anna himself was located and taken prisoner. And the Yellow Rose of Texas, the slave Emily, became famous in legend and song for her role in bringing about the defeat of Santa Anna. Texans still debate whether her "service" that day was prearranged or merely coincidental.

Emily's song has undergone a great deal of change over the past 140 years. One set of original lyrics, not included in the famous Mitch Miller rendition, hailed Emily as "the Maid of Morgan's Point." In place of the chorus line, "She's the sweetest little rose that Texas ever knew," early Texans sang, "She's the sweetest rose of color, this darky ever knew."

SOURCES: Martha Anne Turner, *The Yellow Rose of Texas* (El Paso: Southwestern Studies Monograph–University of Texas, 1971), *passim*; Frank C. Hanighen, *Santa Anna* (New York: Coward-McCann, 1934), p. 88.

★ THE CURIOUS THING ABOUT TEXAS ★

If Texans ever begin worrying that they do not have enough political influence in the country, they have a remedy open to them which is not available to other Americans. By the terms of the treaty annexing Texas to the Union, the state has the right to divide itself at any time into as many as five states. This right gives the state the power to create eight more senators and four more governors.

SOURCE: David Wallechinsky and Irving Wallace, *The People's Almanac #2* (New York: Bantam, 1978), p. 327.

★ THE END OF AN INDIAN TRIBE ★

The Karankawa Indians used to inhabit the lower gulf plains of southern Texas and northern Mexico. But with the arrival of the white man and his diseases and technology, the Karankawas, for all practical purposes, ceased to exist. The tribe met its final demise in the middle of a white man's war, the Texas Revolution of 1836.

Captain Dimmit owned a ranch near the mouth of the Lavaca River, approximately ninety miles north of present-day Corpus Christi. In dealing with the Karankawas, he was always friendly, and was in the habit of giving members of the tribe beef whenever they were in the area. When the Texas Revolution broke out, however, Dimmit left his ranch to serve with the Texans. The Karankawas knew nothing of the war, so when the tribe appeared at Dimmit's ranch and found it deserted, the Indians went out, rounded up a few cattle, and helped themselves to some beef. While eating, a party of Mexican soldiers rode up and demanded to know what the Indians were doing.

"Oh," the Indians innocently replied, "it's all right; we are Captain Dimmit's friends."

When the Mexicans heard this, they charged, killing many and causing the rest to flee. The remaining Karankawas regrouped shortly thereafter, but soon met a party of Americans. Fearing another assault, and not understanding why they had been attacked at Dimmit's ranch, the hapless Indians played it safe with this new group of non-Indians and began shouting, *"Viva Mexico!"*

Immediately, the Americans attacked, and only a few Indians were able to escape to the nearby canebrakes. The two attacks ended the Karankawa tribe, and it was virtually never seen or heard of again in Texas or Mexico.

SOURCE: Noah Smithwick, *The Evolution of a State* (Austin, Texas: Gammel, 1900), pp. 22–23.

★ CACTUS CHASTITY ★

The chastity of young, unmarried, and untouched girls was a matter of great importance and respect among some Indian nations. The girls' protection sometimes included the use of rather unusual methods. The Cheyenne would knot a rope around a young girl's waist and thighs in such a manner that she could not have intercourse without great trouble. The women had to wear the rope at night, or whenever they traveled abroad, from the time they reached puberty until they were married.

Other tribes used more drastic measures. Anthropologists have unearthed in a rock shelter in Val Verde County, Texas, a prickly-pear cactus (*Opuntia*) leaf that most likely had served as the southwestern equivalent of a chastity belt. The cactus leaf curves inward slightly at its base. On the concave side there are no spines or bristles. There are no long bristles on the other side, but there are clumps of small, sharp bristles. Two-thirds of the cactus's circumference is reinforced by two strands of slender bear grass, with loops sewn in at the base and the two sides. With these loops an Indian girl could attach the cactus to her sides by using a girdle-type garment. The cactus then formed a considerable deterrent to would-be seducers.

SOURCE: Melvin R. Gilmore, "An Interesting Vegetal Artifact from the Pecos Region of Texas," *University of Texas Bulletin: Anthropological Papers*, I (September 8, 1937), 21.

★ HENRY CLAY SAVES THE COMMON LAW ★

"When Henry Clay was young, and a brilliant member of the legislature of Kentucky, one of the old Buckskins heard him quote the old common law of England as decisive of the case then under discussion. The old fellow was astonished and, jumping up, began, 'Mr. Speaker, I want to know, sir, if what that gentleman said is true? Are we all livin' under English common law?' The speaker informed the anxious inquirer that the common law was recognized as part of the law of the land. 'Well, sir,' resumed Buckskin, 'when I remember our fathers, and some of us, fit, bled and died to be free of English law, I don't want to be under any of it any longer. And I make a motion that it be repealed right away.' The motion was seconded. The Kentucky blood was up. The Buckskins fired off speech after speech, and Mr. Clay had as much as he could do to explain the matter and save the legislature of Kentucky from repealing the common law of England."

SOURCE: *Harper's New Monthly Magazine*, February 1858, p. 425.

★ HENRY CLAY SPEAKS FOR THE ★ PRESENT GENERATION

"Uniformly cheerful while on the floor, [Clay] sometimes indulged in repartee. The late General Alexander Smyth, of Virginia, a man of ability and research, was an excessively tedious speaker, worrying the House and prolonging his speeches by numerous quotations. On one of these occasions, when he had been more than ordinarily tiresome, while hunting up an authority, he observed to Mr. Clay, who was sitting near him, 'You, sir, speak for the present generation, but I speak for posterity.' 'Yes,' said Mr. Clay, 'and you seem resolved to speak until the arrival of your audience.'"

SOURCE: Epes Sargent, *The Life and Public Services of Henry Clay* (Auburn, N.Y.: Derby & Miller, 1852), p. 111.

★ JOHN VAN BUREN'S PROPHECY ★

John Van Buren, speaking to his father, the President: "You think you're going to be remembered because you were president; but you are going to be remembered because you are the father of John Van Buren."

★ THE VICE PRESIDENT WHO SOLD HIS ★ MISTRESS AT AUCTION

Fidelity to a single woman was not a conspicuous trait of Richard Mentor Johnson, vice president under Martin Van Buren. The Kentuckian never married and ran through at least three women during his life. Undoubtedly, he would have led a conventional married life if his mother had allowed him to wed a New England schoolteacher with whom he fell madly in love as a young man. But she didn't, and her interference marked Johnson permanently. He promised that in time she would regret her action.

When Johnson's father died a few years later, the irate son took his revenge. As part of his inheritance he received a female slave, Julia Chinn, whom his mother had raised as a child. Johnson decided to take Chinn as his mistress. He put her in charge of his home, introduced her into society as his wife, and had several children by her. When he was elected to the United States Senate, he took Chinn to Washington and referred to her in public as his wife. When people refused to accept her at parties, Johnson himself virtually ceased going out.

Chinn died during a cholera epidemic in Kentucky in 1833. But Johnson was not prepared to spend the rest of his life as a bachelor. Although he was in the national spotlight and was even mentioned as a possible presidential candidate, he took other black slaves as mistresses. Even after he was elected vice president by the Senate in 1837, he continued to have black concubines.

Johnson was not without a sense of personal morality, however. Despite his blatant exploitation of female slaves (which was rather ironic considering that he was a champion of the working classes), Johnson still believed in a few old-fashioned virtues.

When one of his mistresses proved unfaithful, the Vice President put her up for sale and then took her sister as his next mistress.

SOURCE: George Stimpson, *A Book about American Politics* (New York: Harper & Brothers, 1952), p. 133.

★ MYTH OF THE LOG CABIN ★

William Henry Harrison was not born in a log cabin. The son of one of the signers of the Declaration of Independence, he was actually born in a two-and-a-half-story mansion, made of red brick, located on a large plantation on the James River. The myth that he was born in a cabin came about during the presidential election of 1840, when a Democratic newspaper charged that Harrison (who was a Whig) wanted little more from life than a pension, plenty of hard cider, and a log cabin. The Whigs promptly turned the taunt to their advantage and presented Harrison as the candidate of humble origins. The tactic was so successful that Whig Daniel Webster actually apologized once to a crowd for *not* being born in a log cabin. He added, however, that his elder brothers and sisters had been. Harrison himself never actually said that he was born in a log cabin, but neither did he say that he was not. And he did make references to his "log cabin home," though he really lived in an ordinary house. The only log cabin he could ever legitimately claim was one that happened to be on his property but which he certainly never lived in.

SOURCE: Clinton Weslager, *The Log Cabin* (New Brunswick, N.J.: Rutgers University Press, 1969), p. 273.

★ A MOST PROLIFIC PRESIDENT ★

On April 15, 1815, Mary Tyler, the first child of future president John Tyler, was born. On June 13, 1860, forty-five years later, Tyler's last child, Pearl, was born. All totaled, Tyler fathered fifteen children by two wives. No U.S. president sired more offspring.

Tyler maried his first wife, Letitia Christian, in 1813. Twenty-nine years and eight children later, Letitia became the first first lady to die in the White House. Tyler became the first president to marry while in office, approximately two years later, when he married his second wife, Julia Gardiner. Julia, born in 1820, was five years younger than Tyler's oldest child. Together, the president and his second wife had seven children. When the fifth one arrived in 1853, Tyler proudly admitted he was "not likely to let the [family] name become extinct." Tyler died in 1862 at age seventy-one. Julia Tyler lived until 1889, twenty-seven years after her husband and forty-five years after her presidential marriage. She died at age sixty-nine.

SOURCE: Robert Seager, *And Tyler Too* (New York: McGraw-Hill, 1963), pp. 102, 359.

★ JOHN TYLER'S FAITHFUL HORSE ★

Appearing on a grave in Virginia is the following inscription written by President John Tyler:

"Here lies the body of my good horse, 'The General.' For twenty years he bore me around the circuit of my practice, and in all that time he never made a blunder. Would that his master could say the same!"

SOURCE: Joseph Kane, *Facts about the Presidents*, 2d ed. (New York: H. W. Wilson, 1968), p. 74.

★ PRESIDENTIAL ADVICE FOR APPOINTMENTS ★

In 1845, President James K. Polk selected James Buchanan as his secretary of state. When former president Andrew Jackson heard of the appointment, he was incensed. "But General," complained Polk, "you yourself appointed him minister to Russia in your first term [as president]." "Yes, I did," replied Old Hickory. "It was as far as I could send him out of my sight and where he could do the least harm! I would have sent him to the North Pole if we had kept a minister there."

SOURCE: A. C. Buell, *History of Andrew Jackson* (New York: Scribner's, 1904), p. 404.

★ THE MOST SUCCESSFUL PRESIDENT ★

Judged by his ability to keep his promises, James K. Polk was the most successful president in American history. During the 1844 election, candidate Polk made five major promises: to acquire California from Mexico, to settle the Oregon dispute, to lower the tariff, to establish a sub-treasury, and to retire from the office after four years. When Polk left office, his campaign promises had all been fulfilled.

SOURCE: Thomas Bailey, *The American Pageant*, 4th ed. (Lexington, Mass.: Heath, 1971), pp. 306–7.

★ A HISTORICAL SOLUTION TO A ★ RECENT PROBLEM

In late March 1974, millions and millions of blackbirds, deviating from their normal migratory pattern, descended upon the area around Graceham, Maryland. A disaster of epic proportions precipitated, with national news coverage of citizens employing everything from firecrackers to electronic sound devices to move the birds along. It was not the area's first encounter with hordes of blackbirds.

In the middle 1800s another multitude of blackbirds landed on the farm of Dr. Fredric Dorsey in nearby Washington County, Maryland. Dorsey soon became so exasperated he scattered wheat soaked in arsenic over his fields. To wash the foreign substance from their throats, the poisoned birds rushed to the stream which passed through Dorsey's farm—where millions of the blackbirds quickly dropped dead. By the next morning the congestion of dead birds had completely dammed up the stream and put over a quarter mile of Dorsey's farm under water. The birds had caused more damage dead than alive.

SOURCE: Thomas J. C. Williams, *A History of Washington County, Maryland* (1906; rpt. Baltimore: Regional, 1968), p. 267.

★ PROSTITUTES IN THE THEATER ★

Promoters often wonder what the best way is to fill a theater. In the first half of the nineteenth century free admission for "women of infamy" was the favorite method of assuring a full house. With a sizable number of prostitutes inside, the paying customers were not far behind. And with paying customers came high ticket receipts for the theater.

Seating for the "ladies" was limited to the upper gallery of the auditorium. This arrangement affected not only the rowdiness but the design and construction of the house. Shouts and noises emanating from the upper gallery often bore little relation to the action on the stage. Back staircases leading directly to the upper gallery began to be designed for many theaters, assuring that good citizens merely out for the show would not have to enter or leave by the same door as those going to work. This door also provided an easy exit in case any theatergoers decided they were no longer interested in watching the show.

If receipts were low, managers would sometimes send messengers directly to a house of prostitution to distribute block tickets. This type of publicity attracted even more girls than usual and brought scores of men eager to pay money to get into the theater.

Occasionally, however, public pressure forced a theater manager to close the upper gallery. Almost without fail, this reduced patronage and profits. Many theatergoers were simply not interested in seeing only the show that occurred onstage. At Boston's Tremont Theater, after one such interdiction, "scarcely fifty persons were present."

Churchmen and "respectable" members of society despised the theater and the upper gallery. According to the Reverend Phineas Densmore Gurley, President Lincoln's assassination by the actor John Wilkes Booth was God's way of showing Americans the evil character and influence of the theater.

SOURCE: Claudia Johnson, "That Guilty Third Tier: Prostitution in Nineteenth-Century American Theaters," *American Quarterly*, XXVII (December 1975), 581.

★ DIVORCE, AMERICAN STYLE ★

When Ruthey Ann Hansley of Hanover County, North Carolina, sued her husband for divorce in March of 1845, the state supreme court revealed how difficult marriage dissolutions could be in mid-century.

Ruthey Ann married Samuel G. Hansley in 1836. According to later divorce proceedings, the couple lived together for years "enjoying much happiness." Yet, for reasons unknown to Mrs. Hansley, her husband suddenly began drinking and was often intoxicated for periods of up to a month. Samuel then began committing a great number of "outrages against the modesty and decency" of his wife, including cruel beatings and all-night absences from their bed. Ruthey Ann next discovered that her husband had been cohabiting with a black slave woman, Lucy. Yet, as was reasonable for any middle-nineteenth-century wife, she "tried to endure."

Samuel then became bolder. He abandoned Ruthey Ann's bed completely. Lucy moved into the house and was given responsibility for the running of the home. Labeling his wife an "incumbrance," Samuel openly and repeatedly ordered her to give place to the slave Lucy, and encouraged the slave to treat his wife with disdain and disrespect.

But Samuel's ill treatment had just started. He began starving his wife for two to three days, and at other times literally threw her out of the house and locked the doors behind her. There she would be compelled to remain for an entire night, unprotected from the weather. Samuel's cruelest brutality occurred when he repeatedly compelled Ruthey Ann to sleep in the same bed with him and Lucy. Once in bed, Lucy was treated "as his wife." Ruthey Ann was too afraid to object. Finally, she could bear no more and fled to the home of her brother, in August 1844. Six months later Ruthey Ann filed for divorce.

The Superior Court of Hanover County granted Ruthey Ann a divorce from Samuel and began an inquiry into a proper financial settlement. Samuel appealed the divorce to the North Carolina Supreme Court, probably to try to avoid a severe alimony ruling. In Raleigh, the high court concluded that although Ruthey Ann's treatment had been unfortunate, and that Samuel's conduct certainly constituted grounds for divorce, it could not be proved that an adulterous relationship between Samuel and Lucy still existed. The major witness to this relationship, Ruthey Ann, had left in August! No further evidence of continued adultery had been introduced, save an off-the-cuff comment by Samuel that he would sooner sell everything he owned than part with Lucy. Therefore, a marriage reconciliation was not beyond hope. The lower court's ruling was overturned and the divorce denied.

SOURCE: Willie Lee Rose, ed., *A Documentary History of Slavery in America* (New York: Oxford University Press, 1976), p. 428.

★ THE SENATOR ELECTED BY A MEXICAN BULLET ★

"In 1848, [James Shields] was elected to the United States Senate [from Illinois], succeeding Senator Breese, who was a candidate for re-election. At the battle of Cerro Gordo, in the war against Mexico, he was shot through the lungs, the ball passing out at his back. His nomination over a man so distinguished as Judge Breese was a surprise to many, and was the reward for his gallantry and wound. His political enemies said his recovery was marvelous, and that his wound was miraculously cured, so that no scar could be seen where the bullet entered and passed out of his body, all of which was untrue. The morning after the nomination, Mr. Butterfield, who was as violent a whig as General Shields was a democrat, met one of the judges in the Supreme Court room, who expressed his astonishment at the result, but added the judge, 'It was the war and that Mexican bullet that did the business.' 'Yes,' answered Mr. Butterfield dryly, 'and what a wonderful shot that was! The ball went clean through Shields without hurting him, or even leaving a scar, and killed Breese a thousand miles away.'"

James Shields is the only man in American history who ever represented three different states in the U.S. Senate: Illinois for six years, Minnesota for one year, and Missouri for thirty-nine days. Remarkably, Shields was an Irish immigrant.

. Arnold, *The Life of Abraham Lincoln* (Chicago: A. C.
& Company, 1918), p. 58n.

★ TWELVE STROKES FOR BAD COOKING ★

Following is an actual list of some of the punishments meted out
to sailors in the United States Navy in 1848:

For bad cooking	12 strokes of the whip
For stealing a major's wig	12 strokes of the whip
For skulking	12 strokes of the whip
For running into debt on shore	12 strokes of the whip
For tearing a sailor's frock	9 strokes of the whip
For filthiness	12 strokes of the whip
For striking a schoolmaster	12 strokes of the whip
For drunkenness and breaking into the liquor closet	12 strokes of the whip
For noise at quarters	6 strokes of the whip
For bad language	12 strokes of the whip
For dirty and unwashed clothes	12 strokes of the whip
For being out of hammock after hours	12 strokes of the whip
For throwing overboard the top of a spittoon	6 strokes of the whip
For taking bread out of oven	9 strokes of the whip
For neglecting mess utensils	12 strokes of the whip
For taking clothes on shore to sell	12 strokes of the whip
For skylarking (running up and down the rigging of a ship)	6 strokes of the whip
For being naked on deck	9 strokes of the whip

SOURCE: Horace Greeley, ed., *The Tribune Almanac for the Years 1838 to
1868* (New York: New York Tribune, 1868), I. 37.

★ THE FORGOTTEN PRESIDENT ★

His gravestone reads, "President of U.S. one day." His name? David Rice Atchison.

Atchison may have been president of the United States for one day, but no one is sure. The facts are these. Atchison was president pro tempore of the Senate on March 4, 1849, the day President James K. Polk's term expired at noon and one day before Zachary Taylor was sworn in (Taylor refused to take the oath on March 4, since that was a Sunday). Because Polk's vice president had resigned a few days before, Atchison, it would seem, was technically the only person legally allowed to exercise the powers of the presidency—by virtue of his being third in line in the succession. According to the law, the president pro tempore automatically became president when the presidency and vice presidency were vacant.

Nothing happened during Atchison's one day in office, though a few friends jokingly requested appointments to the cabinet. Atchison later told the St. Louis *Globe-Democrat:* "I went to bed. There had been two or three busy nights finishing up the work of the Senate, and I slept most of that Sunday."

SOURCE: David Wallechinsky and Irving Wallace, *The People's Almanac # 2* (New York: Bantam, 1978), pp. 178–80.

★ THE RELATION OF TWO PRESIDENTS ★

Jefferson Davis, the first and only president of the Confederacy, was once married to the daughter of President Zachary Taylor. In 1832, at Fort Crawford, in present-day Wisconsin, Davis met and fell in love with Sarah Knox Taylor. He was twenty-three and fresh out of West Point; she was eighteen. Sarah Knox's father, future president Zachary Taylor, was the post's commandant.

"Knox," as she was called by family and friends, loved Davis as much as he loved her. But Zachary Taylor saw only ill fortune for the couple. He was a military officer, and because of this his wife had spent a good part of her life in isolated army forts. He did not wish to see his daughter face the same type of existence. Legend has it that Taylor also disliked Davis for personal reasons. But whatever the reason, Taylor absolutely denied the young couple his permission to marry. Instead, they waited two and a half years, until Knox was of legal age. Davis then resigned from the army, presumably to prevent any interference Taylor might have attempted through military channels, and Jeff and Knox married with the full knowledge—and full disapproval—of Zachary Taylor.

Before the newlyweds reached the home on Davis's Mississippi cotton plantation, however, they traveled to Louisiana, where they visited some of Davis's relatives whom Knox had never met. There, near St. Francisville, both bride and groom contracted malarial fever. Davis recovered, but Sarah Knox, after waiting two and a half years to marry, died in her husband's arms on September 15, 1835, just three months to the day after her wedding. Brokenhearted, Jefferson Davis traveled awhile to regain his health, then became a total recluse at Brierfield, his Mississippi plantation. For the next eight years he saw virtually no one except his slaves and his brother.

In the middle 1840s, Davis began to reappear in public. He remarried in 1845 and was elected to the House of Representatives from Mississippi.

When the Mexican War broke out, he was chosen to lead a regiment of Mississippi volunteers. His commanding officer during a large part of the war was General Zachary Taylor, his former father-in-law. Davis and his regiment fought gallantly at the Battle of Monterey, and Davis himself was wounded at the Battle of Buena Vista. The night after he was wounded, General Taylor appeared at Davis's tent and told the Mississippian, "My daughter was a better judge of men than I was."

Zachary Taylor, a Whig, became president in 1849. Jefferson Davis was a United States senator from Mississippi and the recognized leader of the Southern Democrats. Politically, Davis and Taylor were on opposite sides of the fence, Davis edging toward secessionism, Taylor threatening to lead the army personally against any state that declared its independence from the Union. Persons taken in rebellion, Taylor promised, he would hang— more speedily than he had hanged deserters and spies in Mexico.

Yet, personally, Davis and his second wife were treated as part of the Taylor family. Throughout the general's short term as president, the Davises were in and out of the White House almost as constantly as if Knox had never died.

SOURCE: Holman Hamilton, *The Three Kentucky Presidents* (Lexington, Ky.: University of Kentucky Press, 1978), pp. 12–13.

★ TEN CENTS TOO MUCH TO PAY TO ★ BE PRESIDENT

Some people would spend every cent they had to be nominated for the U.S. presidency. Not Zachary Taylor. When the Whig party nominated him as their presidential candidate in early June 1848, their letter officially notifying him of the nomination carried no postage. When it reached Taylor's home, the future president refused to pay the ten cents' postage due. It was not until July that Taylor learned he was the Whig candidate.

Actually, the Post Office had issued its first stamp only one year before Taylor's nomination, in 1847. Before that time, and continuing for a time afterwards, mail was paid for by the recipient. Taylor, one of America's great heroes in the Mexican War, received volumes of postage-due mail from all across the country. Rather than spend a small fortune on unsolicited mail from total strangers, the hero of the Battle of Buena Vista routinely refused most of his mail. Thus, when the Whig presidential nomination letter came, it was turned away unread. Taylor, of course, won the presidency in November 1848. He was the last Whig ever to be elected president.

SOURCE: Joseph Kane, *Facts about the Presidents*, 2d ed. (New York: H. W. Wilson, 1968), p. 83.

★ THE TRUE VALUE OF THE MEXICAN WAR ★

Following the Mexican War, President Zachary Taylor commissioned Captain William Tecumseh Sherman to explore and survey the newly acquired lands of New Mexico, Arizona, and California. For two years Sherman traversed the sandy, cactus-ridden areas, then returned to Washington. Later, the future Civil War hero called at the White House.

"Well, Captain," inquired the President, "will [the new possessions] pay for the blood and treasure spent in the war?"

"Well, General," replied Sherman, "it cost us one hundred millions of dollars and ten thousand men to carry to the war with Mexico."

"Yes, fully that," returned a satisfied Taylor, "but we got Arizona, New Mexico, and Southern California."

"Well, General," reminisced Sherman about the arid land he had just left, "I've been out there and looked them over,—all that country,—and between you and me I feel that we'll have to go to war again. Yes, we've got to have another war."

"What for?" asked the surprised President.

"Why," answered the captain, "to make 'em take the darn country back!"

SOURCE: Melville D. Landon, *Eli Perkins: Thirty Years of Wit* (New York: Cassell, 1891), p. 32.

BILLY YANK AND JOHNNY REB

★ 〰

"There they are, cutting each other's throats, because one half of them prefer hiring their servants for life, and the other by the hour."

—THOMAS CARLYLE

★ SCRAPBOOK OF THE TIMES ★

- In the wake of the discovery of gold in California, sailors in July 1850 deserted 500 ships in San Francisco to seek their fortunes.
- Harriet Beecher Stowe received torrents of abuse from Southerners for writing *Uncle Tom's Cabin*. She also received a black man's ear.
- Eighteen fifty-one was the first year a Christmas tree was put in an American church—in Cleveland, Ohio.
- While president, Franklin Pierce was arrested after accidentally running down an old woman with his horse. He was released after the arresting officer discovered the identity of the prisoner.
- In 1856, Secretary of War Jefferson Davis ordered seventy camels brought to the United States. He wanted the camels to provide transportation for military personnel pending the completion of the transcontinental railroad. The camels were used for about two years with some success.
- James Buchanan, the only bachelor president, almost married. It was in 1819, when he became engaged to a Miss Anne Caroline Coleman. Buchanan apparently loved his fiancée dearly, but she broke off the engagement and a week later died, probably by suicide. During the rest of his life Buchanan maintained an ironclad silence about his relationship with Miss Coleman.

- The noted abolitionist, Wendell Phillips, commenting on politicians in 1860: "You can always get the truth from an American statesman after he has turned seventy, or given up all hope of the presidency."
- The first time anyone used the temporary-insanity defense in a trial in the United States was in 1859. The defendant, Congressman Dan Sickles, was accused of murdering his wife's paramour. There was no doubt in anyone's mind that Sickles had indeed killed his wife's lover, but the jury acquitted him on the grounds of temporary insanity.
- The first shot of the Civil War was fired by Edmund Ruffin at Fort Sumter. Ruffin was a leading sectionalist who, at the end of the war, committed suicide rather than face the defeat of the Confederacy.
- One of the correspondents of the New York *Tribune* during the Civil War was Karl Marx, who reported on politics in Europe.
- Julia Ward Howe sold her "Battle Hymn of the Republic" to the *Atlantic Monthly* in 1862 for five dollars.
- In late 1862, General Grant issued the following order: "The Jews, as a class violating every regulation of trade established by the Treasury Department and also Department orders, are hereby expelled from the department within twenty-four hours." It has been said, in Grant's defense, that by "Jews" he simply meant peddlers and traders, and that he was not anti-Semitic.
- By 1861 there were only two countries in the Western world other than the United States which maintained slavery: Cuba and Brazil.
- General Stonewall Jackson sent this message to the Confederate War Department: "Send me more men and fewer questions."
- The last veteran of the American Revolution died in 1867.
- The expression, "waving the bloody shirt," meaning the desire to stir up the passions of the Civil War, came into use in 1867 after Senator Ben Butler stood up during the trial of Andrew Johnson and waved a shirt stained with blood.

- Republican Senator Benjamin Wade voted to convict Johnson though he would have become president if Johnson had been found guilty. Johnson was acquitted by only one vote.
- The first racially mixed jury in the United States was impaneled after the Civil War to judge Jefferson Davis. Davis was allowed to go free before the trial began, however.
- The first Negro elected to the U.S. Senate was Hiram Revels of Mississippi. Ironically, Revels's seat had last been filled by Jefferson Davis.
- During his presidency U. S. Grant put many relatives on the federal payroll, including his father, as postmaster at Covington, Kentucky; a brother-in-law, as minister to Denmark; another brother-in-law, as appraiser of customs in San Francisco; and still another brother-in-law as collector of the Port of New Orleans. In all, Grant gave federal positions to thirteen relatives.
- While president, Grant was arrested for speeding in his horse carriage.
- In 1872, Congress passed a law requiring officials of both houses to deprive members of a day's salary for every day's absence, except in the case of illness. The law has been enforced only twice since it was passed.
- Congress in 1873 gave itself a salary raise of 50 percent and made it retroactive for two years.
- The ice-cream soda was invented by accident in 1874, when Robert M. Green ran out of sweet cream and substituted vanilla ice cream in sodas he was selling at the semicentennial celebration of the Franklin Institute in Philadelphia.
- The life expectancy of Americans in 1876 was about forty.
- Custer's march on the Little Big Horn, which cost him his life as well as the lives of 224 other men, was carried out against the orders of his superiors.
- William Cullen Bryant's last words were: "Whose house is this? What street are we in? Why did you bring me here?"
- "Anybody," Jay Gould once remarked, "can make a fortune. It takes genius to hold on to one."

★ DYSAESTHESIA AETHIOPICA: ★
WHY SLAVES ARE LAZY

A question that puzzled slaveowners throughout history was why their slaves had poor work habits. Seemingly obvious answers, such as the absence of incentives and cruel conditions, were often ignored. One of the more interesting theories of slave misbehavior was put forth by Dr. Samuel W. Cartwright of Louisiana, who attributed slave misconduct to a disease, Dysaesthesia Aethiopica, which overseers erroneously labeled "rascality."

The sickness caused slaves to "do much mischief" which often appeared "as if intentional." They became destructive and wasteful in their work. A slave suffering from DA would perform his tasks "in a headlong, careless manner, treading down with his feet or cutting with his hoe the plants" he was trying to cultivate. The symptoms, reported Dr. Cartwright, all reflected "the stupidness of mind and insensibility of the nerves induced by the disease."

The doctor also attributed deviate slave behavior to the Negro's poor sleeping habits. "In bed," he explained, "when disposing themselves for sleep, the young and old, male and female, instinctively cover their heads and faces, as if to insure the inhalation of warm, impure air, loaded with carbonic acid and acqueous vapor. The natural effect of this practice is imperfect atmospherization of the blood— one of the heaviest chains that binds the negro to slavery."

SOURCES: Kenneth M. Stampp, *The Peculiar Institution* (New York: Vintage, 1956), pp. 102–3; Samuel Cartwright, "Diseases of Negroes," *De Bow's Review*, O.S. XI (August 1851), 210.

★ IN THE NAME OF SCIENCE ★

Slavery could be a brutal institution. In the middle 1800s, Mr. Stevens, a large slaveowner from Georgia, fell seriously ill and was cared for by Dr. Hamilton. As a gesture of gratitude, Stevens told the doctor that if he ever needed a favor, his former patient would do anything to help. In conjunction with his medical practice, Dr. Hamilton researched cures for sunstroke. In Stevens's kind offer he found a monumental opportunity to further his studies. Rather than simply make hypotheses about potential cures, Dr. Hamilton borrowed one John Brown, a field slave who had belonged to Stevens for fourteen years. With John as the subject, Dr. Hamilton would now be able to compare his remedies firsthand. Of course, to keep John from missing his daily work in the fields, the experiments had to be conducted at night.

First the doctor ordered a hole dug three and a half feet deep, by three feet in length, by two and a half feet wide. Into the pit went fired, dried red oak bark, with the embers removed. Across the bottom the doctor put a plank with a stool. Brown was then

A slave auction as pictured in the Northern press. (*Harper's Weekly*, January 12, 1867, p. 24.)

forced to strip and sit on the stool, while wet blankets were placed over the hole, locking in the heat. Hamilton allowed a slight gap in the blankets for Brown's head. After half an hour John Brown fainted. Dr. Hamilton then carefully measured the heat in the hole.

Every three or four days Dr. Hamilton repeated the experiment, each time with John taking one of the doctor's sunstroke remedies before descending into the hole. The time required for John to faint and the heat of the hole were meticulously recorded. Finally Dr. Hamilton concluded that cayenne pepper tea, a common household remedy at the time, afforded the best protection against heat. But every household had easy access to cayenne pepper. So Dr. Hamilton marketed pills containing ordinary flour, which, when dissolved in cayenne pepper tea—the instruction indicated—would give a person stamina and fortitude against sunstroke. Reportedly, Hamilton made a large fortune from the sale of his pills.

But the doctor was not through experimenting with John Brown. Next he tried to discover how deep John's black skin went, which left the slave with lifelong scars on his legs, hands, and feet. In all, Dr. Hamilton used John as a human guinea pig for approximately nine months. When the experiments made John too weak to work in the fields, he was transferred to the slave carpenter crew. John soon "got a liking for" this work and, in the end, claimed to be much happier swinging a hammer than he had been swinging his hoe in the fields.

SOURCE: John F. Bayliss, ed., *Black Slave Narratives* (New York: Collier, 1973), pp. 77–80.

★ FREEDOM BY MAIL ★

Escapes from slavery were difficult. Henry "Box" Brown, a slave in Richmond, Virginia, often prayed to God about escaping. Finally, according to Brown's later narrative, instructions came from up above to "go and get a box and put yourself in it." Brown wasted no time. He had the plantation carpenter construct a box the same size as the largest boxes commonly shipped in those days. He then poked three small gimlet holes in the three-foot-by-two-foot crate and carefully marked "this side up with care" on the outside.

With only a "bladder" of water, Brown placed himself inside.

The slave soon learned that travel in a small box could be try-ing even though the trip had been sanctioned by the Almighty. At the express office the box was thrown in a corner upside down. Soon Brown and the box were loaded onto the baggage car, where, with good fortune, the crate happened to fall right side up. Next in his travel northward toward freedom, Brown was transferred to a steamboat, and again placed on his head. For what Brown later estimated was about an hour and a half, he rode upside down on his head. In a short time his "eyes were almost swollen out of their sockets, and the veins on [his] temple seemed ready to burst." His arms and hands were so numb he could not move them. The thought of calling for help, if anyone was even in a position to hear, became more and more appealing than the prospect of dying a slow, gruesome death in the box. Freedom, however, remained Brown's goal, and he resigned himself to either achieving it or dying. Finally, Brown was taken off the boat and placed in a wagon. After a rough ride, the box was thrown down so hard that the runaway's neck was almost broken. But Brown had reached his destination, freedom and Philadelphia.

He was ecstatic. After the group of Northerners he had been mailed to removed the box lid, the former slave promptly stood up and fainted.

Brown was lucky. Another slave, a girl, who attempted to mail herself to freedom contracted "brain fever" and emerged from her box gray-headed. For the rest of her life she appeared ten years older than she actually was.

SOURCE: John F. Bayliss, ed., *Black Slave Narratives* (New York: Collier, 1973), pp. 191–96.

★ WENDELL PHILLIPS AND THE SLAVE ★

"Before Wendell Phillips, the great Abolitionist, was very well known, he had occasion to visit Charleston, South Carolina, and put up at a hotel. In the morning he ordered his breakfast served in his room, and was waited upon by a slave.

"Mr. Phillips seized upon the opportunity to impress upon the Negro, in a sentimental way, that he regarded him as a man and brother, and more than that, he himself was for the abolition of slavery.

"The Negro, however, seemed more anxious about his patron's breakfast than he was about his own position in the social scale or the conditions of his soul, until finally Mr. Phillips became discouraged and told the servant to go away, saying that he could not bear to be waited upon by a slave.

"'You must excuse me,' said the Negro, 'I am obliged to stay here 'cause I'm responsible for the silverware.'"

SOURCE: *Negro Digest*, June 1946, 72.

★ ON THE YAZOO RIVER ★

In the middle of the nineteenth century Stephen Foster came close to using the name Yazoo in a song. He was writing the words to a "plantation song" and wanted the name of a southern river that had a melodious sound. Since he had never been farther south than Kentucky (he was a native of Pittsburgh), he consulted a map and happened upon the river Yazoo, a tributary of the Mississippi. He liked the name and decided to use it, but when the song was first sung his brother objected that the sound of "Yazoo" wasn't quite right. So Foster dropped it and found another river. In 1851 he published his song: "Swanee River."

SOURCE: Frank Smith, *The Yazoo River* (New York: Rinehart, 1954), p. 10.

★ HORACE GREELEY NEVER SAID IT ★

Next to a few lines from the Declaration of Independence and Lincoln's Gettysburg Address the most quoted remark from American history may very well be Horace Greeley's "Go west, young man, go west." Historians routinely trot out the admonition, as do politicians and columnists. It is known to children almost from the time they begin to read. Remarkably, the line is almost never recalled without mentioning the name of Horace Greeley. But Horace Greeley never said it. The author of the quotation was actually John L. Soule, a little-known Indiana journalist, who published it in the Terre Haute *Express* in 1851. Greeley repeatedly denied that he had said it, and even reprinted the article in which Soule used the expression, but to no avail.

SOURCE: Bergen Evans, ed., *Dictionary of Quotations* (New York: Delacorte Press, 1968), p. 745.

★ FRANKLIN PIERCE PREPARES FOR OFFICE ★

Franklin Pierce entered the office of president with possibly the most devastating personal problems of any chief executive.

Pierce's wife, whom he loved deeply, had never taken pride in her husband's political career. In 1836, while Pierce was a young congressman from New Hampshire, she confided to a close friend, "Oh, how I wish [Franklin] was out of political life! How much better it would be for him on every count!" She was disconsolate when her husband was elected to the Senate, and began not accompanying him to the disagreeable capital when he had to attend Congress. Instead, she remained home in New Hampshire while Pierce roomed alone in a boardinghouse.

The Pierces had lost two children in infancy, but when a third, Benjamin, arrived in 1841, Franklin resigned his Senate seat, returned to New Hampshire, and permanently abandoned further political aspirations. In 1846 he declined a cabinet position in James K. Polk's administration because of feelings for his wife and family. Never again, he told Polk, would he be voluntarily separated from his family for any considerable length of time.

In 1852, Pierce was named as New Hampshire's favorite-son candidate for the Democratic party. No one seriously thought he stood a chance of receiving the nomination. But as the convention became deadlocked, the delegates began looking around for a dark horse. Pierce's name was mentioned, and on the forty-ninth ballot he received the top position on the Democratic ticket.

Jeanie Pierce was not enthralled by the political turn of

events. Benjamin was now eleven years old, and both father and mother literally lived for their only surviving child. Franklin told his wife he had not wanted the nomination, but since the party had drafted him anyway he had to run, for Bennie's sake as well as the nation's. Would not Bennie look fine growing up in the White House? Jeanie Pierce acquiesced, and her husband was elected in November.

After the election, however, Mrs. Pierce discovered that her husband had actually sought the Democratic nomination and had lied to her about his "passive" activities. Then, on January 6, 1853, the Pierces personally witnessed the brutal death of their only living son, Benjamin, in a train accident. In two months Jeanie Pierce had lost her faith in her husband's integrity and her son. The President-elect's relationship with his beloved wife irrevocably and painfully deteriorated. Exhausted and depressed over the loss of his son and the alienation of his wife, Pierce, at that time the youngest man ever elected to the presidency, was inaugurated, on March 4, 1853.

SOURCE: Roy Franklin Nichols, *Franklin Pierce* (Philadelphia: University of Pennsylvania Press, 1958), pp. 535–36.

★ THE MOST ANONYMOUS VICE PRESIDENT ★ OF THEM ALL

William King, elected vice president under Franklin Pierce in 1852, was the only man in history not given the chance to prove the harmlessness of his particular office. A mere forty-five days after he took the oath—in Cuba, the only man to be sworn in as vice president in a foreign country—he died, on April 18, 1853. His death came so quickly that he never even had the opportunity of presiding over the Senate, the only job prescribed for vice presidents by the Constitution.

SOURCE: Joseph Kane, *Facts about the Presidents*, 2d ed. (New York: H. W. Wilson, 1968), p. 94.

★ THE COMMODORE KEEPS HIS PROMISE ★

Cornelius Vanderbilt, the great shipping and railroad magnate, was a man of his word in an era of freewheeling finance. In 1849 the commodore founded a steamship line which ran from the east coast to California. The new line, designed to take advantage of increased traffic resulting from the gold rush of '49, cut the usual steamship fare in half, saved the traveler six hundred miles by crossing land at Nicaragua rather than Panama, and offered dependable service, something unheard of in those days. Profits? The commodore earned over a million dollars on the business venture in the first year.

Yet in 1853, Vanderbilt, diverting his assets to other investments, sold a large block of his stock in the line to a group of Americans known as the Nicaragua Transit Company. The commodore was soon in trouble with his new partners, however, as the Transit group refused to pay him for the stock. Legal prosecution, Vanderbilt knew, would mean a drawn-out, national affair. To avoid all of this, he sent the following note:

"Gentlemen: You have undertaken to cheat me. I won't sue you, for law is too slow. I will ruin you. Yours truly, Cornelius Vanderbilt."

Vanderbilt was not merely making an idle threat. Shortly thereafter he established a new shipping company to compete with the other one on the east coast–California route. In two years he put the Nicaragua Transit Company out of business. All totaled, the commodore spent nine years running ships from the east coast to California, making profits estimated at no less than ten million dollars.

SOURCE: N. S. B. Gras and Henrietta M. Larson, *Casebook in American Business History* (New York: F. S. Crofts, 1939), p. 363.

★ HOW JAY GOULD ACCUMULATED HIS ★
FIRST FORTUNE

In 1853 seventeen-year-old Jay Gould set out for New York to sell an invention which he was sure would make a fortune and revolutionize the world. The invention was a better mousetrap. His grandfather had designed it but given Jay the right to exploit it commercially. The future financier leaped at the opportunity, with big dollar signs racing before his eyes, keeping him awake at night.

That summer he set out for the financial hub of the country to try to interest a small manufacturer in his "can't miss" invention. But when he arrived in New York, disaster struck. As he was riding on the Sixth Avenue trolley, gazing lazily from the back of the car at the impressive tall buildings, a thief took off with the mousetrap, which was sitting on a seat enclosed in an invitingly exquisite mahogany case. When Gould returned to his seat, he immediately called to a conductor, "What has become of my box?"

"Was it yours?" the conductor innocently asked him. "Why, a man who got out and turned down the last street carried it off. If you run, you will probably catch him."

Gould promptly jumped off the car, ran madly back to the last stop, and there caught a glimpse of a big, strong fellow carrying away the mahogany box. Gould shot after the thief, wrestled him to the ground, and grabbed the invention.

Just then a policeman appeared. Immediately the thief accused Gould of trying to steal the mahogany box. Gould attempted to explain that the box was his, but to no avail. It was down to the station for both men. There Gould convinced the police that the box was his by revealing its unusual contents. The thief had no idea what he had stolen, of course, and was thunderstruck to learn that all that expensive-looking box contained was a crude-looking mousetrap.

The next day the New York *Herald* devoted half a column to the incident. It was the first time Jay Gould made it into the papers, for once as a victim.

Within three years Gould succeeded in turning a profit on his mousetrap. By the age of twenty, from the trap and other inventions, he had accumulated the vast savings of $5,000, which in the days before the Civil War was a small fortune.

SOURCES: Matthew Josephson, *The Robber Barons* (New York: Harcourt, Brace, 1934), pp. 38–39; Robert I. Warshaw, *Jay Gould* (New York:

Greenberg, 1928), pp. 42–43; Richard O'Connor, *Gould's Millions* (Garden City, N.Y.: Doubleday, 1962), p. 30.

★ CONDENSING THE WORLD ★

Gail Borden, the inventor of condensed milk and founder of the company that bears his name, aspired to condense much more than milk. "I mean," he once remarked, "to put a potato into a pill box, a pumpkin into a tablespoon, the biggest sort of watermelon into a saucer. . . . The Turks made acres of roses into attar of roses. . . . I intend to make attar of everything."

SOURCE: Daniel Boorstin, *The Americans: The Democratic Experience* (New York: Random House, 1973), p. 313.

★ HISTORY OF BLACK GOLD ★

When an eastern Kentucky salt well suddenly filled up with oil in 1818, becoming the first well in America to produce crude, it was promptly abandoned as useless. Years earlier, Seneca Indians and General Benjamin Lincoln's Revolutionary soldiers had recognized the value of oil for medicinal purposes, but businessmen had not yet discovered its commercial possibilities. When oil was found in more salt wells in the late 1830s, these too were abandoned.

In the 1840s, after many uses for oil had been discovered, businessmen couldn't find enough oil-producing salt wells to supply their needs. These wells were in especially great demand because the only other known way of obtaining crude was by collecting it from springs, creeks, and ditches—a slow and costly method. Incredibly, though everyone knew that salt wells had produced oil, no one thought of drilling a well solely for the purpose of getting oil. When James M. Townsend, president of the Pennsylvania Rock Oil Company, suggested to a friend in the late 1850s that oil could perhaps be produced from a well just like water, he was told, "Nonsense! You're crazy."

In 1857, Edwin L. Drake, an agent for the Rock Oil Company,

became convinced that oil wells could be drilled. Forming his own company in 1859, he commissioned a salt borer to drill him a hole. The man thought Drake was insane, however, and never appeared. Finally Drake hired William A. "Uncle Billy" Smith, an experienced salt borer and blacksmith.

Drilling began in June. Progress was slow, and only about three feet a day were dug. By August 27 the hole was just sixty-nine and one-half feet deep. The next day was Sunday, a rest day. On Monday, Uncle Billy went to the well to begin work and discovered it full of oil. "What's that?" Drake asked him. "That," Uncle Billy responded, "is your fortune!"

It was not to be, however. Drake went to Wall Street, speculated in oil stocks, and went bankrupt. Eventually the state of Pennsylvania granted him an annual pension of $1,500 for his pioneer work in oil.

SOURCE: Daniel Boorstin, *The Americans: The Democratic Experience* (New York: Random House, 1973), pp. 41–46.

★ PRESIDENTS START AS INDENTURED SERVANTS ★

As boys both Millard Fillmore and Andrew Johnson were indentured servants. In this early form of contract labor, the master, for all intents and purposes, owned the servant for the length of his contract (usually five to seven years). The rights of an individual servant were in many respects comparable to the rights of a slave, which were few, of course.

Fillmore and Johnson did not enjoy this type of servitude. Andrew Johnson ran away. The tailor he was indentured to placed an advertisement in the Raleigh, North Carolina, *Gazette*, offering a ten-dollar reward for the capture and return of the future president. Unfortunately for the tailor, Johnson was never caught.

Fillmore was indentured to a clothmaker. After serving his master for several years, he purchased his freedom for thirty dollars.

SOURCE: Sid Frank, *The Presidents* (Maplewood, N.J.: Hammond, 1975), p. 33.

★ FILLMORE REFUSES AN HONORARY DEGREE ★

Fillmore, on refusing to accept an honorary degree of Doctor of Civil Law from Oxford University: "I had not the advantage of a classical education and no man should in my judgment accept a degree he cannot read." In 1846, though he never attended college, the modest Fillmore accepted the chancellorship of the University of Buffalo.

SOURCE: Joseph Kane, *Facts About the Presidents*, 2d ed. (New York: H. W. Wilson, 1968), p. 88.

★ ANDREW JOHNSON'S CHALLENGE TO ★
AN ASSASSIN

Like his mentor and fellow Tennessean, Andrew Jackson, Andrew Johnson did not scare easily. Brought up in the Carolinas, he wore the rough-hewn characteristics of the frontier conspicuously. Election to political office did not soften his manner. When threats were made against his life in 1855 during his campaign for a second term as governor of Tennessee, Johnson challenged his would-be assassins to meet him face to face. "Fellow citizens," he began a speech at one campaign stop, after laying a pistol on the table in front of him, "I have been informed that part of the business to be transacted on the present occasion is the assassination of the individual who now has the honor of addressing you. I beg respectfully to propose that this be the first business in order. Therefore if any man has come here tonight for the purpose indicated, I do not say to him let him speak, but let him shoot."

SOURCE: Joseph Kane, *Facts about the Presidents*, 2d ed. (New York: H. W. Wilson, 1968), p. 118.

★ STUDENTS TRY TO MURDER ★
STONEWALL JACKSON

Stonewall Jackson was a great general, but a horrible teacher. Between his heroic service in the Mexican War and the Civil War,

Jackson taught mathematics at Virginia Military Institute. VMI students hated Jackson because he was stubborn, narrow-minded, and made excessive demands on them. Complaints about the war hero were not always verbal. Once, as Jackson walked near the campus barracks, a couple of particularly vengeful VMI students dropped a brick on him from a third-story window. The brick brushed Jackson's hat, but had it landed on him he very likely would have been killed. Jackson walked straight ahead and did not stop to look up or around.

As fate would have it, when the Civil War began, many VMI students who had formerly hated Professor Jackson served gallantly for the Confederacy under General Jackson.

SOURCE: Frank F. Vandiver, *Mighty Stonewall* (New York: McGraw-Hill, 1957), p. 79.

★ HEAVEN CAN WAIT ★

In the nineteenth century the danger of being buried alive was very great. The mother of Robert E. Lee reportedly was buried alive during one of her catatonic trances. Luckily, a sexton heard noises and scratching from inside the coffin as it was being covered with dirt, and she escaped. Mrs. Lee went on to live several more years. Throughout the century horrible incidents like this came to the attention of the public, prompting the invention of devices to prevent premature burial. One man patented a coffin which allowed a person inside the box to ring a bell aboveground. Another man was actually buried in a coffin with ventilators and a hose that connected him to the outside world.

SOURCE: Mary Cable, *American Manners and Morals* (New York: American Heritage Publishing Company, 1969), p. 175.

★ JOHN BROWN GETS AWAY WITH MURDER ★

In 1858, John Brown was a wanted man. The abolitionist leader, who had killed five proslavery settlers at Potawatomie, Kansas, in 1856, had once again violated the law: at Fort Scott, Kansas, where he was involved in the murder of a storekeeper and the theft of $7,000 worth of goods; and at two plantations in Missouri, where he freed eleven slaves, kidnapped two whites, and made off with a large amount of valuable property.

But Brown was not arrested and was not sent to jail. Astonishingly, he was allowed to remain free, though his crimes were widely condemned—by Southerners, Free Soilers, and an overwhelming majority of Kansans. President Buchanan put a bounty on Brown's head, Missouri's governor made some threats about apprehending him, and the governor of Kansas asked the state legislature to take immediate action against the agitator.

But neither federal nor state authorities dared to arrest Brown. Despite his crimes, he was popular in the North and could not be taken without risking a riot. Abolitionists would not allow him to be captured.

Beyond the arm of the law, "Old Ossawatomie" traveled in the open, repeatedly made speeches in public, and even brazenly

Trial of John Brown in Charlestown, Virginia. (*Harper's Weekly*, November 21, 1859, p. 728.)

appealed for funds to support future attacks on the "peculiar insti-tution." At Grinnell, Iowa, he made two speeches and was cheered by throngs of appreciative townspeople, who provided him with money and supplies. In Chicago he roamed the city at liberty and received $500 that had been raised for him by world-famous detective Allan Pinkerton. In Cleveland he gave several lectures and in Rochester and Boston was treated as a hero.

Brown remained at large throughout 1859 while he collected money and volunteers for fresh attacks on slavery. Finally, after the failure of his notorious raid on Harpers Ferry, in October, he was captured and convicted. On December 2 he was hanged—almost a year after he had engaged in murder, kidnapping, and larceny.

SOURCE: Allan Nevins, *The Emergence of Lincoln* (New York: Scribner's, 1950), II, 23–27.

★ ALEXANDER STEPHENS'S BIG BRAIN ★ AND SMALL BODY

"Alexander H. Stephens never weighed a hundred pounds. Once a burly Georgian got angry at him and said, 'I have half a mind to swal-low you alive.' Stephens retorted, in his high voice, 'If you do, you'll have more brains in your belly than you ever had in your head.'"

SOURCE: Champ Clark, *My Quarter Century of American Politics* (New York: Harper & Brothers, 1920), II, 183.

★ ADVICE ON BECOMING PRESIDENT ★

Before he was nominated for president by the Democratic party in 1860, Stephen A. Douglas delivered an emotion-charged speech on the floor of the Senate denouncing "nigger-worshippers." That evening William Seward, who would later serve as Lincoln's secre-tary of state, walked home with Douglas from the Capitol. Know-ing Douglas's burning desire to receive his party's nomination and to be elected president, Seward offered his friend some advice.

"Douglas," he explained, "no man will ever be president of the United States who spells negro with two g's."

SOURCE: Charles Shriner, ed., *Wit, Wisdom, and Foibles of the Great* (New York: Funk & Wagnalls, 1918), p. 616.

★ LINCOLN SELLS A DRINK TO DOUGLAS ★

"On one occasion [Stephen] Douglas sneeringly referred to the fact that he once saw Lincoln retailing whisky.

"'Yes,' replied Lincoln, 'it is true that the first time I saw Judge Douglas I was selling whisky by the drink. I was on the inside of the bar, and the judge was on outside; I busy selling, he busy buying.'"

SOURCE: Champ Clark, *My Quarter Century of American Politics* (New York: Harper & Brothers, 1920), II, 192.

★ LINCOLN FORGETS A COMMAND ★

"I remember [Lincoln] narrating his first experience in drilling his company [in the Blackhawk War]. He was marching with a front of over twenty men across a field, when he desired to pass through a gateway into the next inclosure.

"'I could not for the life of me,' said he, 'remember the proper word of command for getting my company endwise so that I could get through the gate, so as we came near the gate I shouted, "This company is dismissed for two minutes, when it will fall in again at the other side of the gate!"'"

SOURCE: Ben Perley Poore, "Lincoln," *Reminiscences of Lincoln*, ed. Allen Thorndike Rice (New York: North American Publishing Company, 1886), pp. 218–19.

★ LINCOLN'S MESS ★

A story by Lincoln's law partner, William Herndon:
   "Lincoln had always on the top of our desk a bundle of papers

into which he slipped anything he wished to keep and afterwards refer to. It was a receptacle of general information. Some years ago, on removing the furniture from the office, I took down the bundle and blew from the top the liberal coat of dust that had accumulated thereon. Immediately underneath the string was a slip bearing this endorsement, in his hand: 'When you can't find it anywhere else, look in this.'"

SOURCE: William Herndon, *Life of Lincoln*, ed. Paul M. Angle (Greenwich, Conn.: Fawcett Publications, 1961), p. 263.

★ A FRESHMAN CONGRESSMAN ELECTED ★ SPEAKER OF THE HOUSE

On December 5, 1859, the House of Representatives convened and tried to choose a Speaker. Sixteen men competed, however, and no one was elected. The House adjourned. The next day one of the members, a Missourian, attempted to limit the number of candidates by making taste in books a test of their eligibility. He offered a resolution forbidding anyone who liked Hinton Rowan Helper's *The Impending Crisis*, an antislavery work, to become Speaker. But the resolution was not agreed to. The following day a North Carolinian put forth a resolution requiring every candidate to be against raising the slavery question. But it was not approved either.

Two ballots, five, ten, twenty, thirty, forty . . . and still no Speaker. December passed, then January. The House was immobilized, caught between proslavery and antislavery rivers.

Finally, on February 1, the House elected a Speaker, William Pennington of New Jersey, whose sole qualification seemed to be his lack of enemies. No one knew anything about Pennington—before December 5, 1859, he had not been a member of Congress. The House of Representatives had elected as its Speaker a freshman congressman.

SOURCE: George B. Galloway, *History of the House of Representatives* (New York: Crowell, 1961), pp. 46–47.

★ THE GREATEST SPOILSMAN OF THEM ALL ★

The image of Andrew Jackson throwing out Republicans and replacing them with Democrats when he acceded to the presidency is a strong one. The slogan, "To the victors belong the spoils," is firmly associated with Jackson in the mind of nearly all Americans. But the plain fact is that Jackson did not turn out of office that many people. In all, Old Hickory replaced only 252 out of a total of 612 officers (not including postmasters). Thomas Jefferson, who never earned a reputation as a spoilsman, turned out more incumbents, over half of those in office. But it was Abraham Lincoln who was the preeminent spoilsman. He threw out of office more appointees than any other president in history. His first year in office he replaced 1,457 men, leaving fewer than 200 appointees from previous administrations.

SOURCE: Nathan Miller, *The Founding Finaglers* (New York: David McKay, 1976), pp. 140, 155, 180.

★ JOHN TYLER RETURNS TO CONGRESS ★

When John Tyler retired from the presidency in 1845, his political career did not end. In 1861 he was elected to represent his native state, Virginia, in the House of Representatives—of the Confederacy. He was the only United States president to serve in the rebel government.

SOURCE: Sid Frank, *The Presidents* (Maplewood, N.J.: Hammond, 1975), p. 6.

★ VOLUNTEER AGING WITH A STROKE OF THE PEN ★

In 1861 many sixteen- and seventeen-year-old boys wanted desperately to volunteer for the Union Army and fight for their country. Yet, with what one historian describes as "youthful innocence," these would-be soldiers would not tell an outright lie to their government. The minimum age for a Union soldier was eighteen. Rather than walk into the recruiting office and swear they were eighteen years old, as young volunteers did in other American wars, these young boys would scribble the number "18" on a scrap of paper and place it in the sole of their shoe. Then, when questioned about their age, they could truthfully reply to their government, "I am over 18."

SOURCE: Bruce Catton, *America Goes to War* (Middletown, Conn.: Wesleyan University Press, 1958), p. 49.

★ THE FIRST JEWISH CABINET OFFICER ★

The first Jew to hold an American cabinet office was Judah P. Benjamin of Louisiana. Judged by today's political standards, this observant Jew is one of the most interesting and paradoxical figures in American history.

Benjamin was born in the West Indies in 1811 and raised in Charleston, South Carolina. He graduated from Yale Law School and began practicing law in New Orleans, where he was a state legislator, a founder of the Illinois Central Railroad, and a plantation owner with 140 slaves. Two years after selling his plantation in 1850, he entered the U.S. Senate. He was the second Jew to serve in the Senate.

When Louisiana seceded from the Union in 1861, President Jefferson Davis, who had served in the Senate with Benjamin, appointed him attorney general of the Confederacy. Benjamin later became the Confederacy's secretary of war and then secretary of state. In Varina Davis's autobiography, she reports that Benjamin was her husband's most trusted adviser and that he usually spent twelve hours a day with Davis, shaping every important Confederate tactic and strategy. Benjamin was widely known as "the brains of the Confederacy," although some Southerners blamed him as the war went badly.

In 1864, Benjamin privately persuaded Robert E. Lee and other Confederate military leaders that the South's best chance of victory was to emancipate any slave who volunteered to fight for the Confederacy. Only with this massive influx of new fighting men, Benjamin argued, could independence be won. When the idea became public, it caused a firestorm. "If slaves will make good soldiers," argued Howell Cobb of Georgia against the proposal, "our whole theory of slavery is wrong." After strenuous debate, the idea was rejected as politically untenable. The following year the South was defeated.

When John Wilkes Booth killed Abraham Lincoln in 1865, Davis and Benjamin were suspected of plotting the assassination. In anti-Semitic Northern newspapers, Benjamin was pilloried as Judas while the martyred Lincoln was compared to Christ. When Lee surrendered at Appomattox, Benjamin escaped to England, fearing he would never receive a fair trial if charged with Lincoln's murder.

Benjamin lived out the rest of his life as a barrister in England, where he published a classic legal text on the sale of personal property. Before his death in 1884, he burned his personal papers, leaving historians with little hard evidence of his thoughts and actions. When Benjamin is remembered in American history, it is usually unsympathetically. He is buried in Père-Lachaise Cemetery in Paris, along with Oscar Wilde, Chopin, Marcel Proust, Jim Morrison, Edith Piaf, and others.

The first Jew elected to the U.S. Senate was also from the antebellum South, David Levy Yulee of Florida. Levy County, Florida, is named in his honor. He was first elected in 1845 and, like Benjamin, withdrew from the Senate in 1861 when Florida seceded from the Union. Yulee was a member of the Confederate Congress until 1865. It was not until 1898 that a non-Confederate state elected a Jew to the U.S. Senate, when Joseph Simon of Oregon took office.

SOURCES: Eli N. Evans, *Judah P. Benjamin: The Jewish Confederate* (New York: Free Press, 1988); Kurt F. Stone, *The Congressional Minyan: The Jews of Capitol Hill* (Hoboken, N.J.: KTAV Publishing, 2000).

★ NEGROES IN THE CIVIL WAR ★

In April 1861 the first Negroes were appointed commissioned officers in the Civil War—by the Confederacy, in Louisiana. By war's end 93,000 blacks served in the Confederate Army. About 100,000 blacks fought in the Union Army, and more than 65,000 were killed. There were 30,000 blacks in the Union Navy, about a quarter of the total number of sailors.

SOURCES: Harry Fleischman, *Let's Be Human* (New York: Oceana Publications, 1960), p. 55; Robert Mullen, *Blacks in America's Wars* (New York: Monad Press, 1973), pp. 22–23, 31.

First Louisiana black troops disembarking at Fort Macombe, Louisiana. (*Harper's Weekly*, February 28, 1863, p. 133.)

★ CONFEDERATES OUTSMART YANKEES ★

"[Confederate Captain John Singleton Mosby] got word that the first breech-loading rifles ever made were coming down (a whole shipment of them) to General Pope commanding the Union forces in Virginia. With these rifles the Union Army would undoubtedly defeat Lee and walk into Richmond. So it was believed. The rifles were carried in a great train to Alexandria, loaded onto wagons, some fifty or sixty of which were being escorted by the Thirteenth Pennsylvania Cavalry. The munitions wagon train was to park right in the heart of General Pope's army. Getting this information, Mosby decided to do something about it. It was a very, very wet, rainy, cold spring night. He waited and the escort came along. Mosby, debouching from a road at the left, rode up with the Colonel boot-to-boot. The Colonel saluted. 'Colonel Graham,' said Mosby, 'the orders have been changed; your men are tired and have been thirteen hours on the march. My orders are to take over and escort the wagon train to another parking place.' The Colonel gave over, and Mosby took over, and parked the munitions train in the heart of General Lee's army. Thus it was that the Confederates had breech-loading rifles and ammunition before the Union troops had them."

SOURCE: Louis Brownlow, *The Anatomy of the Anecdote* (Chicago: University of Chicago Press, 1960), pp. 52–53. Reprinted by permission of the University of Chicago Press.

★ THE PROPER WAY TO SURRENDER ★

On the night of September 17, 1862, the most unusual surrender in all of American military history occurred. Braxton Bragg, the Confederate major general, was steadily advancing northward into Kentucky. At Munfordville, where the railroad to Louisville crossed the Green River, Bragg's forces encountered a federal strongpoint of 4,000 men under the command of Colonel John T. Wilder. Bragg's advance guard attacked the Union fortifications twice and were repulsed both times, with moderate losses. Bragg then brought up the rest of his army, completely surrounding the Union position. He sent in a demand for surrender, pointing out that the Yankee case was hopeless.

Late that night, under a white flag of truce, Colonel Wilder crossed the Confederate lines. Wilder asked for a conference with Major General Buckner, a Confederate known to him "not only [as] a professional soldier but [as] an honest gentleman." In Buckner's tent Wilder, who had been an unassuming Indiana businessman until the outbreak of the war, was frank. He was not a military man at all. Yet he wanted to do the right thing. Was it his duty, under the rules of the game, to surrender his badly outnumbered forces or to fight it out?

Buckner was straightforward and to the point. As he later explained, he "would not have deceived that man under those circumstances for anything." The Union soldiers were hemmed in by six times their own number. The Rebels had enough artillery in line to demolish Wilder's position in a few hours. But, on the other hand, Buckner advised, Wilder should fight it out if he thought the federal cause would be helped by the sacrifice of every man. Finally, Buckner took Wilder to see Major General Bragg.

Bragg was curt. Like Buckner, he too would not tell the Union commander what course to follow. But he did want Wilder to make an educated decision. Together the Confederate major general and the Yankee colonel went to inspect the Southern artillery placements. From there Bragg allowed Wilder to begin counting the number of cannon pointed toward the Union fort. Wilder did not need to count them all to realize that his position was hopeless.

Later that day he surrendered.

SOURCE: Bruce Catton, *Terrible Swift Sword* (Garden City, N.Y.: Doubleday, 1963), p. 411.

★ JEFFERSON DAVIS ASSAULTED ★

"The president was returning with Mrs. Davis from one of the customary festivities on a flag of truce boat that had come up the James; walking the street in the night, unattended by his staff, he had to pass the front of Libby prison, where a sentinel paced, and, according to his orders, forced passengers from the sidewalk to the middle of the street. As the president, with his wife on his arm, approached him he ordered him off the pavement. 'I am the president,' replied Mr. Davis; 'allow us to pass.' 'None of your gammon,' replied the soldier, bringing his musket to his shoulder; 'if you don't get into the street I'll blow the top of your head off.' 'But I am Jefferson Davis, man; I am your president—no more of your insolence,' and the president pressed forward. He was rudely thrust back and in a moment had drawn a sword or dagger concealed in his cane and was about to rush on the insolent sentinel when Mrs. Davis flung herself between the strange combatants and by her screams aroused the officer of the guard. Explanations were made and the president went safely home. But, instead of the traditional reward to the faithful sentry that has usually graced such romantic adventures, came an order the next day to degrade the soldier and give him a taste of bread and water for his unwitting insult to the commander-in-chief of the Confederate armies."

SOURCE: Edward A. Pollard, *Life of Jefferson Davis* (Philadelphia: National, 1869), pp. 155–56.

★ MONITOR NEVER BEAT MERRIMAC ★

Despite what nearly everyone believes, there never was a naval battle between two ships named the *Monitor* and the *Merrimac*. On March 9, 1862, there did occur a famous sea battle between two ironclads at Hampton Roads, Virginia; the Northern ship was indeed the celebrated *Monitor*. But the Southern ship was known as the *Virginia*, not the *Merrimac*. Before 1862 the vessel was a wooden steam frigate belonging to the Union and was called the *Merrimac*. But Confederates had seized the ship, after it had been sunk by United States sailors evacuating the Norfolk naval yard, and rechristened it the *Virginia*. The old name continued to be used, however; by Southerners perhaps because it went well with *Monitor*; by Northerners for the same reason and because they probably did not believe the Confederates had any more right to rename a Union ship than they had to secede.

SOURCE: David Donald and James Randall, *The Civil War and Reconstruction*, 2d ed., rev. (Boston: Little, Brown, 1969), pp. 439–44.

★ THE PRUDENT GENERAL ★

"[Confederate General Albert Sidney] Johnston's deliberateness is illustrated by his remark to a precipitate friend who was about to run across the street in front of a carriage driving rapidly, 'There is more room behind that carriage than in front of it.'"

SOURCE: Charles Shriner, ed., *Wit, Wisdom, and Foibles of the Great* (New York: Funk & Wagnalls, 1918), p. 312.

★ RECRUITS AT EASE ★

A group of North Carolina recruits on picket duty near Manassas one evening in 1863 found themselves in one of the less conflict-ridden areas of the Civil War. With not a Yankee within twenty miles, the only thing they had to worry about was an inspection scheduled the next day. One of the new recruits was dismantling his gun to clean and shine it for the upcoming inspection when suddenly General Barham rode up.

"What are you doing there?" asked the general.

"Oh, I am only a kind of a sentinel. Who are you anyhow?"

"Oh, I am only a kind of a brigadier general."

"Hold on," exclaimed the startled recruit, "wait until I get this darned old gun together and I will give you a kind of a present arms."

SOURCE: Melville D. Landon, *Eli Perkins: Thirty Years of Wit* (New York: Cassell, 1891), p. 264.

★ THE MAN WHO GAVE US THE ★ WORD *SIDEBURNS*: CIVIL WAR GENERAL AMBROSE BURNSIDE

(*Leslie's Illustrated*, October 1, 1881, p. 69.)

★ WHEN THE MEDAL OF HONOR ★ COULD BE HAD CHEAPLY

In the spring of 1863 the North seemed to be losing the Civil War. All that year victory after victory had gone to the Confederacy.

The situation would change shortly with the successes at Gettysburg. But in May few had real confidence in the ability of the Union to win the war.

The seriousness of the North's position was dramatically illustrated by an extreme offer Secretary of War Edwin Stanton was forced to make to the 27th Maine Regiment. A week before Gettysburg the members of the 27th were scheduled to leave the army. But Stanton needed them to help defend the capital. Despairing that the men would not reenlist, Stanton, with the approval of Lincoln, promised that the government would award the Medal of Honor to any member of the regiment who reenlisted. Shortly thereafter, simply for reenlisting, 864 members of the 27th Maine Regiment received the prestigious medal.

In 1917 a committee called the Adverse Action Medal of Honor Board took up the cases of the 864 members of the 27th. After some consideration, the board decided that not one of the Maine soldiers should have received the famous award. The board argued that reenlisting in the army was not action "above and beyond the call of duty." And with that the board disqualified every single medal.

SOURCE: *American Heritage*, October-November, 1978, p. 112.

★ THE DIFFERENCE BETWEEN AMERICANS ★
AND ENGLISHMEN

Henry Ward Beecher, the famous preacher-Republican-abolitionist, traveled to England during the Civil War in an attempt to rally British sympathy for the Northern cause. In Manchester, England, he talked for an hour in front of a howling mob of Rebel supporters until he finally got their attention. He was again interrupted by one irate Englishman who assailed the lengthy war. "Why didn't you whip the Confederates in sixty days, as you said you would?"

"Because," snapped Beecher, "we found we had Americans to fight instead of Englishmen."

SOURCE: Thomas Masson, ed., *Little Masterpieces of American Wit and Humor* (New York: Doubleday, Page, 1904), II, 32.

★ HE PROMISED NOT TO SWEAR ★

"One Sabbath, Rev. Dr. Nelson, of the First Presbyterian Church, was preaching earnestly upon the necessity of a pure life, exhorting the men [in Colonel Clinton Fisk's regiment] to beware of the vices incident to the camp, and especially warning them against profanity. The Doctor related the incident of the Commodore who, whenever recruits reported to his vessel for duty, was in the habit of entering into an agreement with them that he should do all the swearing for that vessel: and appealed to the thousand Missouri soldiers in Colonel Fisk's regiment to enter into a solemn covenant that day with the Colonel that he should do all the swearing for the Thirty-third Missouri. The regiment rose to their feet as one man and entered into the covenant. It was a grand spectacle.

"For several months no swearing was heard in the regiment. Col. Fisk became a Brigadier, and followed Price into Arkansas. But one evening [in February 1863] as he sat in front of his head-quarters at Helena, he heard someone down in the bottom-lands near the river, swearing in the most approved Flanders style. On taking observation he discovered that the swearer was a teamster from his own headquarters, a member of his covenanting regi-ment, and a confidential old friend. He was hauling a heavy load of forage from the depôt to camp; his six mules had become rebel-lious with their overload, had run the wagon against a stump and snapped off the pole. The teamster opened his great batteries of wrath and profanity against the mules, the wagon, the Arkansas mud, the Rebels, and Jeff Davis. In the course of an hour after-wards, as the teamster was passing headquarters, the General called to him and said, 'John, did I not hear some one swearing most terribly an hour ago down on the bottom?'

"'I think you did, General.'

"'Do you know who it was?'

"'Yes, sir; it was me, General.'

"'Do you not remember the covenant entered into at Benton Barracks, St. Louis, with Rev. Dr. Nelson, that *I* should do all the swearing for our old regiment?'

"'To be sure I do, General,' said John; 'but then you were not there to do it, *and it had to be done right off!*'"

SOURCE: Edward P. Smith, *Incidents of the United States Christian Commission* (Philadelphia: Lippincott, 1869), pp. 88–89.

★ THE MAN HOOKERS WERE NAMED AFTER: ★
BRIGADIER GENERAL JOE HOOKER

(*Harper's Weekly*, June 5, 1862, p. 421.)

The word "hooker" goes back before the Civil War, to the time the Dutch seaport Hook became famous for its streetwalkers. But not until the War between the States did the term become popular. At that time prostitutes south of Washington, D.C.'s Constitution Avenue began being referred to as Hooker's Division—in honor of Joe Hooker, the Union's preeminent paramour.

SOURCE: William Morris and Mary Morris, *Dictionary of Word and Phrase Origins* (New York: Harper & Row, 1971), p. 290.

★ LINCOLN'S DOUBTS ABOUT ABUTMENTS ★
ON THE SOUTHERN SIDE

One of Lincoln's stories:

"I once knew a good, sound churchman, whom we'll call Brown, who was on a committee to erect a bridge over a very dangerous and rapid river. Architect after architect failed, and at last Brown said he had a friend named Jones, who had built several bridges, and could build this. 'Let's have him in,' said the committee. In came Jones. 'Can you build this bridge, sir?' 'Yes,' replied Jones; 'I could build a bridge to the infernal regions, if necessary.' The sober committee were horrified; but when Jones retired, Brown thought it but fair to defend his friend. 'I know Jones so well,' said he, 'and he is so honest a man, and so good an architect, that, if he states soberly and positively that he can build a bridge to Hades—why, I believe it. But I have my doubts about the abutment on the infernal side.' So [Lincoln added], when politicians said they could harmonize the Northern and Southern wings of the Democracy, why, I believed them. But I had my doubts about the abutment on the Southern side."

SOURCE: Frank Moore, ed., *The Civil War in Song and Story* (New York: Collier, 1889), p. 32.

★ THADDEUS STEVENS'S HOT STOVE ★

"During the American Civil War Thaddeus Stevens warned Lincoln that Simon Cameron was not trustworthy as head of the War Department. 'You don't mean to say you think Cameron would steal?' asked Lincoln. 'No,' was the reply, 'I don't think he'd steal a red-hot stove.' Amused by this, Lincoln repeated it to Cameron, who insisted that Stevens should retract. Going to see Lincoln at the White House, Stevens said, 'Why did you tell Cameron what I said?'—'I thought it was a good joke: I never thought it would make him mad.' 'Well, he is mad and made me promise to retract. So I will. I believe I told you he would *not* steal a red-hot stove. I now take that back.'"

Source: Daniel George, ed., *A Book of Anecdotes* (n.p.: Hulton Press, 1957), p. 308.

★ Lincoln Keeps Petitioners Waiting ★

"In the purlieus of the Capitol at Washington, the story goes that, after the death of Chief Justice [Roger] Taney, and before the appointment of Mr. [Salmon] Chase in his stead, a committee of citizens from the Philadelphia Union League, with a distinguished journalist at their head as chairman, proceeded to Washington, for the purpose of laying before the President the reason why, in their opinion, Mr. Chase should be appointed to the vacancy on the bench. They took with them a memorial addressed to the President, which was read to him by one of the committee. After listening to the memorial, the President said to them, in a very deliberate manner: 'Will you do me the favor to leave that paper with me? I want it in order that, if I appoint Mr. Chase, I may show the friends of the other persons for whom the office is solicited, by how powerful an influence, and by what strong personal recommendations, the claims of Mr. Chase were supported.'

"The committee listened with great satisfaction, and were about to depart, thinking that Mr. Chase was sure of the appointment, when they perceived that Mr. Lincoln had not finished what he intended to say. 'And I want the paper, also,' continued he, after a pause, 'in order that, if I should appoint any other person, I may show his friends how powerful an influence, and what strong recommendations, I was obliged to disregard in appointing him.' The committee departed as wise as they came."

Source: Frank Moore, ed., *The Civil War in Song and Story* (New York: Collier, 1889), p. 440.

★ Tarred and Feathered ★

When Lincoln was asked how he liked being president, he said: "You have heard the story, haven't you, about the man who was tarred and feathered and carried out of town on a rail? A man in the crowd asked him how he liked it. His reply was that if it was not for the honor of the thing, he would rather walk."

SOURCE: Bill Adler, *Presidential Wit* (New York: Trident Press, 1966), pp. 62–63.

★ SOUTHERN GUNS MARKED FOR THE NORTH ★

"General [Philip Henry] Sheridan [a Northerner] bagged two-thirds of his enemy's force and most of the enemy's artillery. In the previous summer . . . , as General Early kept losing gun after gun, great efforts were made to re-supply his losses by sending up fresh guns from Richmond. Upon one of these guns some wag of a Confederate soldier had chalked, 'General Sheridan, care of General Early.'"

SOURCE: Charles Shriner, ed., *Wit, Wisdom, and Foibles of the Great* (New York: Funk & Wagnalls, 1918), p. 566.

★ CONFEDERATES LEARN HOW TO USE MAGGOTS ★

At the Union camp in Chattanooga there were scores of wounded Federal soldiers and Confederate prisoners, but a dearth of medical supplies. Especially scarce were chloroform and lint, which were used to keep maggots out of open wounds. Of course, the chloroform was made available only to Union doctors; Confederate doctors were given nothing. Inevitably, the soldiers in gray became infested with maggots.

But a strange thing happened. Johnny Reb healed faster than Billy Yank. Even the rooms where the Southerners were housed smelled fresher and seemed healthier than the Yankee sickrooms.

Unwittingly, the Southern doctors had stumbled onto a great discovery: maggots can be useful in stopping the growth of bacteria and in keeping open wounds clean. A French surgeon had learned this in the Napoleonic Wars, but his finding had been ignored. History repeated itself at Chattanooga; Union doctors, disbelieving the obvious, continued to treat patients with chloroform.

SOURCE: Rudolph Marx, *Health of the Presidents* (New York: Putnam, 1960), p. 225.

★ SHOOTING MAKES 'EM MADDER ★

A reminiscence of the Civil War by Mississippi Congressman John Allen:

"Upon one occasion [a Confederate cavalry colonel] was leading his regiment in one of the most gallant retreats ever engaged in. The Yankees were riding close behind and pressing the boys everywhere. There were some indiscreet men in his command who would turn around and fire at them occasionally. With hat off, from the head of his regiment he turned and looked back and gave this command:

"'Boys, stop that shooting; it just makes 'em madder.'"

SOURCE: U.S. Congress, *Congressional Record* (Washington: Government Printing Office, 1873–), XXV, 1495.

★ FIRST WITH ALMOST THE MOST IN WAR ★

The number-one rule in war is "You have to get there first with the most," and the man who first verbalized it was Confederate General Nathan B. Forrest. But, as the Yankees learned, Forrest had a corollary to his rule. When one simply cannot be first with the most, improvise.

In late September 1864 the Confederate armies were almost uniformly in retreat. General Forrest, however, was leading his troops northward along the railroad from Decatur, Alabama, toward Nashville, Tennessee. Time was the most crucial factor in his attack. To disrupt Union operations elsewhere, Forrest would have to strike with speed into Tennessee.

The strongest Union post between the two cities was at

Athens, Alabama. On the night of September 23, Forrest surrounded this Fort and at seven o'clock the next morning began his attack. But the general was apprehensive. A Union relief column, Confederate intelligence had reported, was already traveling south from Nashville. The Union post at Athens was well manned and well fortified. Forrest would be able to take it only with a great sacrifice of time and men. But the Confederate general had a plan.

After the Union commander, Colonel Wallace Campbell, had rejected a demand for immediate and unconditional surrender, Forrest proposed a personal meeting between the two leaders to prevent further needless bloodshed. Campbell agreed and left his fort to meet Forrest. The Confederate general then accompanied his guest on an inspection of the Rebel troops. Together, they traversed the encircled fort, counting Confederate men and artillery. Each time the visitors left a particular detachment, however, the Rebel soldiers, according to Forrest's prearranged orders, would pack up and move to another position, artillery and all. Pretty soon, Forrest and Campbell would arrive at the new encampment and continue to tally the number of Rebel soldiers and guns. By the time he returned to his fort, Colonel Campbell had visited over eight thousand Southern troops, both cavalry and infantry. All were supplied with ample artillery. Unaware of the advancing relief column, Campbell rode back to his command and exclaimed, "The jig is up; pull down the flag."

Meanwhile, the relief column of seven hundred Michigan and Ohio regulars courageously battled toward Athens for three hours. Finally, within sight of the fort, they were overcome and forced to surrender—just in time to witness the evacuation of the Union fort. On that one day of "battle," Forrest took 1,300 prisoners, plus horses and weapons.

All totaled, General Forrest's army killed or wounded 1,000 Union soldiers and captured 2,360 on its campaign into Tennessee. Of the approximately 4,500 Confederates involved, only 47 were killed and 293 wounded. Despite these great victories, the Confederacy was on its last legs. Atlanta had fallen and Sherman was marching through Georgia. Forrest's trickery had come too late and was too little to help.

SOURCE: Robert Selph Henry, *"First with the Most" Forrest* (Indianapolis: Bobbs-Merrill, 1944), p. 354.

★ ZACK CHANDLER AND THE WOODCHUCK ★

"Zack Chandler [the Republican politician from Michigan] had three men working in a saw-mill in the woods below Saginaw. During Lincoln's last campaign, Zack went up to the saw-mill to see how the men were going to vote. He found that each had a different political faith. One was a Democrat, one a Republican and one a Greenbacker. A farm-boy had just killed a fine woodchuck, and Zack offered to give it to the man who would give the best reason for his political faith.

"'I'm a Republican,' said the first man, 'because my party freed the slave, put down the rebellion and never fired on the old flag.'

"'Good!' said old Zack.

"'And I am a Greenbacker,' said the second man, 'because if my party should get into power every man would have a pocket full of money.'

"'First-rate!' said Uncle Zack. 'And now you,' addressing the third: 'Why are you a Democrat?'

"'Because, sir,' said the man trying to think of a good democratic answer—'because—because I want that woodchuck!'"

SOURCE: Melville D. Landon, *Kings of the Platform and Pulpit* (Chicago: Werner Company, 1895), p. 546.

★ A LIE REVEALED ★

One of the most famous anecdotes from the Civil War was Lincoln's response to complaints about General Grant's drinking habits: "If I knew what brand he used, I'd send every other general in the field a barrel of it." Unfortunately, the story was completely untrue. Lincoln himself said so when someone for once asked him whether he had actually said what was commonly attributed to him.

SOURCE: David Homer Bates, *Lincoln Stories* (New York: William E. Rudge, 1926), pp. 49–50.

★ LINCOLN ORDERS A PASS TO RICHMOND ★

"A gentleman called upon President Lincoln before the fall of Richmond and solicited a pass for that place. 'I should be very happy to oblige you,' said the President, 'if my passes were respected; but the fact is, I have, within the past two years, given passes to two hundred and fifty thousand men to go to Richmond and not one has got there yet.'"

SOURCE: Paul Selby, ed., *Anecdotal Lincoln* (Chicago: Thompson & Thomas, 1900), p. 21.

★ THE CONFEDERACY OFFERS TO ★ ABOLISH SLAVERY

Fearing the imminent collapse of his government, Confederate President Jefferson Davis in March 1865 notified England and France that the South would be willing to abolish slavery in exchange for diplomatic recognition. Before either European power could respond, however, the war was over.

SOURCE: Emory Thomas, *The Confederacy as a Revolutionary Experience* (Englewood Cliffs, N.J.: Prentice-Hall, 1971), pp. 130–31.

★ JEFFERSON DAVIS'S CAPTURE ★

When Jefferson Davis was captured by Federal troops on May 10, 1865, he was wearing his wife's raglan and shawl, which he had put on in the dark of his tent just as he was trying to make an escape. Cartoonists and illustrators in the North mercilessly pictured Davis fleeing in woman's dress.

Jefferson Davis fleeing in woman's clothes. (*Harper's Weekly*, May 27, 1865, p. 336.)

★ BEFORE JOHN WILKES BOOTH ★

The bullet that killed Abraham Lincoln at Ford's Theater in April 1865 was not the first to be fired at the President. Twice before he had been shot at, both times while on his way to the Soldier's Home. In 1861, while riding alone at night to the Home, Lincoln was fired upon by a man standing less than fifty yards away. In August 1864 he was again shot at, the bullet passing through the upper part of his stovepipe hat. In both cases Lincoln joked about the incidents and ordered that they not be publicized.

SOURCE: Carl Sandburg, *Abraham Lincoln* (New York: Harcourt, Brace, 1939), II, 205–6; III, 440–41.

Lincoln in his coffin. (*Harper's Weekly*, May 6, 1865, p. 285.)

★ AMERICAN WAR DEATHS GREATEST IN ★ THE CIVIL WAR

Counting both Northerners and Southerners, more American lives were lost in the Civil War than in any other conflict. Following is a list showing the number of lives lost in every war in United States history. The figures include death owing to sickness:

American Revolution	c. 25,324
War of 1812	c. 2,260
Mexican War	c. 13,283
Civil War	c. 498,332
Union	364,511

Confederacy	c. 133,821
Spanish-American War	2,446
World War I	116,516
World War II	405,399
Korean War	54,246
Vietnam War	56,244
Persian Gulf War	148

SOURCE: *Dictionary of American History,* rev. ed. (New York: Scribner's, 1976), VII, 232.

★ A NEGRO BUYS A PRESIDENT'S PLANTATION ★

Although Jefferson Davis was the president of the Confederacy, at the end of the South's struggle to maintain slavery his brother, Joseph, with the full consent and approval of Jefferson, sold the Davis plantation to Benjamin T. Montgomery, one of their former slaves. When the Civil War ended, Jefferson Davis was thrown into a Union jail, and Brierfield, the Davis plantation, was in bad financial condition. Davis's brother, aged and unable to properly manage the land, sold it to Montgomery in 1866 for $300,000. Four years later Joseph Davis died.

Unfortunately, by 1881 it became clear that Montgomery would not be able to continue making payments on the land. The mortgage was foreclosed and Brierfield reverted to Davis's hands. Ownership of the plantation by the Davis family continued until the early 1950s, when Brierfield became a state museum.

SOURCE: James T. McIntosh, ed., *The Papers of Jefferson Davis* (Baton Rouge: Louisiana State University Press, 1974), II, 245.

★ SANTA CLAUS DID NOT ALWAYS HAVE A BEARD ★

In the seventeenth century Dutch settlers brought the myth of Santa Claus to America. But the Santa Claus they brought did not look anything like the jolly figure pictured today. Their Santa was

Santa Claus as he was pictured before Nast, in 1858. (*Harper's Weekly*, December 25, 1858, p. 817.)

Nast's 1863 Santa Claus. (Albert Bigelow Paine, *Th. Nast: His Period and His Pictures* [New York: Macmillan, 1904], p. 22.)

tall, slender, and very dignified. Around the beginning of the nineteenth century Santa took on the appearance of a jolly figure. In 1809, Washington Irving imagined Santa as a bulky man who smoked a pipe and wore a Dutch broad-brimmed hat and baggy breeches. Later in the century artists pictured Santa as a fat man, with brown hair and a big smile. Finally, in 1863, Thomas Nast drew a picture of Santa as a jolly old man with a white beard and wide girth—the first picture of Santa as he looks today.

SOURCE: Albert Bigelow Paine, *Th. Nast: His Period and His Pictures* (New York: Macmillan, 1904), p. 22.

★ AN IMPEACHMENT DEFENSE ★

Senator William S. Groesbeck made the closing speech at President Andrew Johnson's impeachment trial. With a solemn tone and tears

Andy Johnson. (*Harper's Weekly*, April 15, 1862, p. 221.)

in his eyes, Senator Groesbeck defended Johnson with these final remarks: "The President is not a learned man, like many of you senators; his light is the feeble light of the Constitution."

SOURCE: Melville D. Landon, *Eli Perkins: Thirty Years of Wit* (New York: Cassell, 1891), p. 249.

★ HE GAVE US CHEWING GUM? ★

In the fall of 1866, Mexican general Santa Anna, exiled from his native land, lived on Staten Island, New York. His new interpreter and secretary, a young American named James Adams, noticed how the old general would constantly cut slices from an unknown tropical vegetable and place the pieces in his mouth. Inquiring, Adams learned that the substance was called "chicle." When Santa Anna left New York the following May, the young interpreter persuaded him to leave behind his supply of chicle. Adams then began experimenting with the substance, adding different sweetening agents to bolster the flavor. Soon he had "invented" chewing gum. When Adams introduced his new product to the American public, he found a willing and hungry market. Later, Adams founded the Adams Chewing Gum Company, and Americans, helped by a most unlikely Mexican source, have been chewing gum ever since.

SOURCE: Oakah L. Jones Jr., *Santa Anna* (New York: Twayne, 1968), p. 145.

★ THE CENTURY'S GREATEST INVENTION ★

Sarah Hale, editor for forty years of *Godey's Ladies Magazine*, editorialized in 1868: "There are many great men who go unrewarded for the services they render to humanity: even their names are lost, while we daily bless their inventions. One of these is he (if it was not a lady) who introduced the use of visiting cards."

SOURCE: Sarah Hale, *Manners* (Boston: J. E. Tilton, 1868), p. 217.

★ STEVENS IS CARRIED TO THE CAPITOL ★

"There is nothing finer, as I think, in the annals of humor than [Thaddeus Stevens's] quaint question to David Reese and John Chauncey, the two officers of the House, who in his last days used to carry him in a large armchair from his lodgings across the public grounds up the broad stairs of the noble Capitol[.] 'Who will be so good to me, and take me up in their strong arms [Stevens asked], when you two mighty men are gone?'"

As Stevens lay on his bed, sick and near death, friend John Hickman came for a visit and told the congressman that he looked well. "Ah John," quickly replied Stevens, "it is not my appearance, but my disappearance, that troubles me."

SOURCE: *Harper's New Monthly Magazine*, August 1871, p. 478.

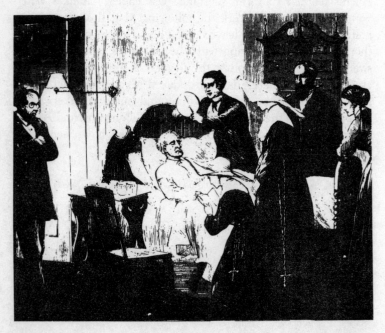

Thaddeus Stevens on his deathbed. (*Harper's Weekly*, August 29, 1868, p. 548.)

★ MISTREATING GOVERNMENT PROPERTY ★

"Ex-Governor Chapman, of Alabama, one of the sturdy old patriots who are honored by the special hatred of the Yankees, suffered seriously in wanton spiteful depredations on his property near Huntsville, Ala. Thinking that some of the doings of the Yankee villains were beyond orders, he waited on the Yankee commander (Colonel Alexander), and stated his case:

"Colonel.—'Well, Governor, I don't think you have any property about here.'

"'Well, sir, if it is not mine, be so kind as to inform me whose it is?'

"Colonel.—'It is the property of the Government of the United States, sir.'

"Governor.—'Ah! very well, Colonel, I have come to inform you, then, that your soldiers are treating the property of the United States Government d——d badly. Good day, Colonel.'"

SOURCE: Nora F. M. Davidson, ed., *Cullings from the Confederacy* (Washington: Rufus H. Darby, 1903), p. 63.

★ THE POST–CIVIL WAR DEATH OF ★ CLEMENT VALLANDIGHAM

Everyone has to die. Yet it seems unfair for a great life to end in the way Ohio Civil War Congressman Clement L. Vallandigham's did.

Vallandigham was a leader of the "Peace Democrats," or "Copperheads," a group of Middle Westerners and Northerners who opposed war as a means of bringing the South back into the Union. As the struggle dragged on, Copperhead cries for peace found increasing support among the war-weary populace. By 1862, the movement's strength began seriously to hinder the administration's military efforts. Finally, to force at least a semblance of unity on what was left of the nation, Lincoln used his power as a wartime commander-in-chief and had Vallandigham arrested and banished to the Confederacy as a traitor. But the Ohio congressman managed to escape and fled to Canada, where he continued to espouse his Copperhead views.

In 1871, however, Vallandigham was back in his native Ohio practicing law. In June he accepted the case of Thomas McGehan, a young rowdy accused of killing one Tom Myers in a barroom brawl. Vallandigham's defense would show that Myers had actually shot himself while drawing a pistol out of his pocket while rising from a kneeling position. On June 16, in a conference with his fellow defense lawyers, Vallandigham outlined the performance he planned to give before the jury the next day. Taking a pistol from his bureau and placing it in his right pocket, the former congressman knelt to exactly the same position he claimed Myers had assumed immediately prior to the shooting. Slowly Vallandigham lifted the pistol out of his pocket as he rose. When the muzzle of the gun cleared his pocket, Vallandigham carefully cocked the weapon and placed it in precisely the same spot he believed Myers had held his gun when it had discharged.

"There, that's the way Myers held it," Vallandigham commented triumphantly, "only he was getting up, not standing erect." Suddenly, however, there was a flash and the muffled sound of a shot. "My God," the former Copperhead leader exclaimed, "I've shot myself!" Twelve hours later Vallandigham died.

His argument must have impressed someone. After a hung jury and a mistrial, a third jury found Thomas McGehan innocent of the charge of murder.

SOURCE: Frank L. Klement, *The Limits of Dissent* (Lexington: University of Kentucky Press, 1970), pp. 309–10.

★ GRANT OVERTURNS DARWIN ★

Henry Adams, a Harvard historian and the descendant of two presidents, would not be expected to have had much sympathy for U. S. Grant. He didn't.

"Simple-minded beyond the experience of Wall Street or State Street, [Grant] resorted, like most men of the same intellectual calibre, to commonplaces when at a loss for expression: 'Let us have peace!' or, 'The best way to treat a bad law is to execute it'; or a score of such reversible sentences generally to be gauged by their sententiousness; but sometimes he made one doubt his good faith; as when he seriously remarked to a particularly bright young woman that *Venice would be a fine city if it were drained*. In Mark Twain, this suggestion would have taken rank among his best witticisms; in Grant it was a measure of simplicity not singular. . . . That, two thousand years after Alexander the Great and Julius Caesar, a man like Grant should be called—and should actually and truly be—the highest product of the most advanced evolution, made evolution ludicrous. One must be as commonplace as Grant's own commonplaces to maintain such an absurdity. The progress of evolution from President Washington to President Grant, was alone evidence enough to upset Darwin."

SOURCE: Henry Adams, *The Education of Henry Adams* (1906; rpt. New York: Modern Library, 1931), pp. 265–66.

★ GRANT LEARNS ABOUT GOLF ★

A friend once tried to get U. S. Grant to learn the increasingly popular game of golf. After persistent prodding, Grant finally consented to go to a course as an observer.

As they arrived, a man stepped up to the ball and began hacking furiously at it with his driver. Dirt and grass flew everywhere, but the beginning golfer simply could not connect with the ball. Somewhat confused, Grant turned to his friend.

"That does look like very good exercise," he admitted. "What is the little white ball for?"

SOURCE: Sid Frank, *The Presidents* (Maplewood, N.J.: Hammond, 1975), p. 48.

★ HORACE GREELEY RUNS LAST IN 1872 ★

Horace Greeley, the famous newspaper editor, was the presidential candidate of the Democratic and Liberal Republican parties in the 1872 contest against Republican incumbent U. S. Grant. Greeley personally and politically despised Grant. He considered the

Republican's Reconstruction policies a sham and thought Grant a poor judge of moral character. But the Civil War general was popular with the American public, and Greeley was given virtually no chance of winning.

To compound his problems, Greeley's wife was on her deathbed. In late October a very despondent Greeley wrote a friend, "I disagree with you about death. I wish it came faster. . . . I wish she were to be laid in her grave next week, and I to follow her the week after." On October 30, Molly Greeley died.

On November 4, with political defeat looming, Greeley wrote the same friend, "I am not dead but I wish I were. My house is desolate, my future dark, my heart a stone." The next day Grant was reelected by a landslide. Greeley carried only six states. Both politically and financially, he was ruined.

By the middle of November, Greeley, his mental condition rapidly deteriorating, was removed to a private sanitarium in upper New York State. On November 29, three weeks after the election, he died.

When the electoral college met in December, the sixty-six Democratic electors who had been elected for Greeley were instructed to use their own judgment in casting their ballots. Most voted for Thomas A. Hendricks of Illinois. Three Georgia electors, however, insisted on voting for the deceased Greeley. Congress refused to count their votes.

Thus, not only did Horace Greeley, in less than a month's time, lose his wife, the election, his money, his mind, and his own life, but he also became the only presidential candidate of a major political party to receive no electoral votes.

SOURCE: Glyndon G. Van Deusen, *Horace Greeley* (Philadelphia: University of Pennsylvania Press, 1953), p. 420.

★ P. T. BARNUM'S BRICK TRICK ★

P. T. Barnum describes how he snared suckers to his New York museum:

"One morning a stout, hearty-looking man, came into my

ticket-office and begged some money. I asked him why he did not work and earn his living? He replied that he could get nothing to do and that he would be glad of any job at a dollar a day. I handed him a quarter of a dollar, told him to go and get his breakfast and return, and I would employ him at light labor at a dollar and a half a day. When he returned I gave him five common bricks.

"'Now,' said I, 'go and lay a brick on the sidewalk at the corner of Broadway and Ann Street; another close by the Museum; a third diagonally across the way at the corner of Broadway and Vesey Street, by the Astor House; put down the fourth on the sidewalk in front of St. Paul's Church, opposite; then, with the fifth brick in hand, take up a rapid march from one point to the other, making the circuit, exchanging your brick at every point, and say nothing to anyone.'

"'What is the object of this?' inquired the man.

"'No matter,' I replied; 'all you need to know is that it brings you fifteen cents wages per hour. It is a bit of my fun, and to assist me properly you must seem to be as deaf as a post; wear a serious countenance; answer no questions; pay no attention to any one; but attend faithfully to the work and at the end of every hour by St. Paul's clock show this ticket at the Museum door; enter, walking solemnly through every hall in the building; pass out, and resume your work.'

"With the remark that it was 'all one to him, so long as he could earn his living,' the man placed his bricks and began his round. Half an hour afterwards, at least five hundred people were watching his mysterious movements. He had assumed a military step and bearing, and looking as sober as a judge, he made no response whatever to the constant inquiries as to the object of his singular conduct. At the end of the first hour, the sidewalks in the vicinity were packed with people all anxious to solve the mystery. The man, as directed, then went into the Museum, devoting fifteen minutes to a solemn survey of the halls, and afterwards returning to his round. This was repeated every hour till sundown and whenever the man went into the Museum a dozen or more persons would buy tickets and follow him, hoping to gratify their curiosity in regard to the purpose of his movements. This was continued for several days—the curious people who followed the man into the Museum considerably more than paying his wages—till finally the policeman, to whom I had imparted my object, complained that the obstruction of the sidewalk by crowds had become so serious that I must call in my 'brick man.'"

SOURCE: P. T. Barnum, *Struggles and Triumphs* (Hartford, Conn.: J. B. Burr, 1869), pp. 121–23.

★ REPUBLICANS AND THE NEGRO ★

A speech by Abram Jasper at a Negro political gathering at Louisville:

"Feller freemen, you all know me. I am Abram Jasper, a republican from way back. When there has been any work to do, I has done it. When there has been any votin' to do, I has voted early and often. When there has been any fightin' to do, I have been in the thick of it. I are 'bove proof, old line and tax paid. And I has seed many changes, too. I has seed the Republicans up. I has seed the Democrats up. But I is yit to see a nigger up. T'other night I had a dream. I dreampt that I died and went to Heaven. When I got to de pearly gates ole Salt Peter he says:

"'Who's dar?' sez he.

"'Abram Jasper,' sez I.

"'Is you mounted or is you afoot?' says he.

"'I is afoot,' says I.

"'Well, you can't get in here,' says he. 'Nobody 'lowed in here 'cept them as come mounted,' says he.

"'Dat's hard on me,' says I, 'arter comin' all dat distance. 'But he never says nothin' mo, and so I starts back an' about half way down de hill who does I meet but dat good ol' Horace Greeley. 'Whar's you gwine, Mr. Greeley?' says I.

"'I is gwine to heaven wid Mr. Sumner,' says he.

"'Why, Horace,' says I, ''tain't no use. I's just been up dar an nobody's 'lowed to get in 'cept dey comes mounted, an' you's afoot.'

"'Is dat so?' says he.

"Mr. Greeley sorter scratched his head, an' arter awhile he

says, says he: 'Abram, I tell what let's do. You is a likely lad. Suppose you git down on all fours and Sumner and I'll mount an' ride you in, an' dat way we kin all git in.'

"'Gen'lemen,' says I, 'do you think you could work it?'

"'I know I kin,' says bof of 'em.

"So down I gits on all fours, and Greeley and Sumner gets astraddle, an' we ambles up de hill agin, an' prances up to de gate, an' old Salt Peter says:

"'Who's dar?'

"'We is, Charles Sumner and Horace Greeley,' shouted Horace.

"'Is you both mounted or is you afoot?' says Peter.

"'We is bof mounted,' says Mr. Greeley.

"'All right,' says Peter, 'all right,' says he; 'jest hitch your hoss outside, gen'lemen, and come right in.'"

SOURCE: Melville D. Landon, *Kings of the Platform and Pulpit* (Chicago: Werner Company, 1895), pp. 558–59.

★ CIVIL WAR GENERAL BENJAMIN ★ WADE'S HUMOR

"Once Wade was crossing the plains. On the train a man said: 'All this region needs is more water and better society.'

"'Yes,' growled 'Old Ben'; 'that's all Hades needs to make it an ideal dwelling-place!'"

SOURCE: Champ Clark, *My Quarter Century of American Politics* (New York: Harper & Brothers, 1920), II, 214.

★ YALE LEARNS ABOUT EVOLUTION ★

Some ideas are better left alone than refuted. Herbert Spencer, the famous English philosopher, adapted the ideas of Darwinian evolution to biology, psychology, sociology, and other fields of study. Many people, of course, did not agree with Spencer's reasoning.

At Yale University, President Noah Porter personally conducted a volunteer class on Spencer's *First Principles*, trying to refute them. By the end of the term, however, every member of the class had become a believer in Social Darwinism.

SOURCE: Richard Hofstadter, *Social Darwinism in American Thought*, rev. ed. (Boston: Beacon Press, 1955), pp. 20–21.

★ LEE SCARES DUNG OUT OF YANKEES ★

"[Zebulon Vance, former Confederate governor of North Carolina,] turned the tables on the Yankees when he went to Massachusetts [after the war, as a senator,] to deliver a lecture. The Bay Staters, knowing his droll manner and practical jokes, baited him by hanging Robert E. Lee's picture in the men's outhouse. When Vance returned from it, he disappointed them by remaining silent. Finally, they were compelled to query him.

"'Senator, did you see General Lee's picture hanging in the privy?' someone asked.

"'Yes,' Vance replied indifferently.

"'Well, what did you think of it?' they prodded.

"'I thought it was very appropriate,' he responded. 'That is a good place for General Lee's picture. If ever a man lived who could scare the dung out of the Yankees, that man was Robert E. Lee!'"

SOURCE: Glen Tucker, *Zeb Vance: Champion of Freedom* (Indianapolis: Bobbs-Merrill, 1965), p. 7.

★ THE WAY IT DIDN'T HAPPEN ★

Years after the end of the Civil War two Confederate veterans were reminiscing about their battles around Paducah, Kentucky, when one of them bragged, "I remember when we pushed those damyankees all the way across the Ohio and up into Illinois."

"I was there," the other man stated sharply, "and I'm afraid that wasn't the way it happened at all. Those Yankees drove *us* out of Paducah and almost to the Tennessee line."

The bragging veteran suddenly took on a dour look and then wryly commented: "Another good story ruined by an eyewitness."

SOURCE: Alben Barkley, *That Reminds Me* (Garden City, N.Y.: Doubleday, 1954), p. 35.

★ THE GREATEST MAN SOUTH CAROLINA ★ HAS EVER SEEN

"One morning in Saratoga Governor Curtin, the old war governor of Pennsylvania, [sat down on a] balcony by Senator Wade Hampton, one of the proudest of the old South Carolina rebels. They [were] both keen wits and both gentlemen of the old school.

"'I tell you, governor,' began General Hampton enthusiastically, 'South Carolina is a great State, sir—a great State.'

"'Yes; South Carolina is a State to be proud of,' said Governor Curtin. 'I agree with you. I knew a good many distinguished people down there myself—and splendid people they were too—as brave as Julius Caesar and as chivalric as the Huguenots.'

"'You did, sir!' said Senator Hampton, warming up with a brotherly sympathy. 'Then you really knew public men who have lived in our old Calhoun State? You knew them?'

"'Oh, bless you, yes!' continued Governor Curtin, drawing his chair up confidently. 'I knew some of the greatest men your State has ever seen—knew them intimately too, sir.'

"'Who did you know down there in our old Palmetto State?' asked Senator Hampton, handing Governor Curtin his cigar to light from.

"'Well, sir, I knew General Sherman, and General Kilpatrick, and—'

"'Great guns!'"

SOURCE: Melville D. Landon, *Eli Perkins: Thirty Years of Wit* (New York: Cassell, 1891), p. 37.

THE GREAT BARBECUE

★ ≋

"An honest politician is one who when he is bought will stay bought."

—SIMON CAMERON

"Claim everything, concede nothing, and if defeated, allege fraud."

—TAMMANY MAXIM

★ SCRAPBOOK OF THE TIMES ★

- Henry Ward Beecher in 1877: "Is the great working class oppressed? Yes, undoubtedly it is. God has intended the great to be great and the little to be little."
- The one machine indispensable to modern capitalism, the cash register, was patented in 1879 by James Ritty.
- James A. Garfield was the only man in U.S. history who was a congressman, a senator-elect, and a president-elect at the same time.
- A short time after assuming the presidency, James Garfield blurted out, "My God! What is there in this place that a man should ever want to get in it?"
- While in prison for the assassination of President Garfield, Charles Guiteau received upwards of a hundred letters and telegrams a day approving his murderous deed.
- During a lecture tour of the United States in 1881, Oxford professor of history Edward A. Freeman repeatedly remarked that "the best remedy for whatever is amiss in America would be if every Irishman should kill a negro and be hanged for it." Freeman later claimed that he had not meant to be insulting.
- *Nouveau riche* extravagances included a dinner held in honor

of a dog who was given a $15,000 diamond collar, and a man who had little holes drilled into his teeth so that he could have a diamond-studded smile.

- The Vanderbilts built a mausoleum at New Dorp, Staten Island, that cost $1,500,000, and was protected by a watchman twenty-four hours a day.

- According to Louis L'Amour, the Western-fiction writer, the red light became associated with prostitution because late-nineteenth-century train conductors who visited whorehouses often left their red lamps hanging outside.

- In 1884 the "latest social craze"—according to numerous advertisements—was displaying framed pictures on walls.

- More than a million shares were first traded in one day on the New York Stock Exchange in 1886.

- There was an advertisement in 1888 for the Bradley Two Wheeler (a carriage), which carried the headline: "Guaranteed Absolutely Free from Horse Motion."

- Polygamy was legal in Utah until the year 1890.

- Electric lights were installed in the White House during the administration of Benjamin Harrison. Harrison and his wife were so afraid of electricity that they left the job of turning the light switches on and off to the servants.

- Before his election to the White House, Grover Cleveland candidly admitted to the public that he had sired a son out of wedlock. His admission led to the jeer: "Ma Ma, where's my Pa/Gone to the White House, Ha, Ha, Ha."

- As a boy Franklin Roosevelt was told by Cleveland: "Franklin, I hope you never become president."

★ THE UNSUSPECTED ORIGINS OF TAMMANY HALL ★

Tammany Hall was not always dominated by Irish immigrants. The organization began in 1789 as a fraternal society of native Americans and was actually formed, in part, to oppose immigrants. The first constitution of the society provided that only whites were eligible for the exalted post of Sachem. When the

Tammany Hall. (*Harper's Weekly*, July 11, 1868, p. 433.)

organization turned to politics in the early 1800s, it refused to endorse Irish candidates for office. As late as 1817, Thomas A. Emmet, the Irish patriot refugee, was unable to win the support of Tammany Hall even after two hundred rowdy Irishmen stormed society headquarters. Not until the 1840s, when a flood of immigrants came from Ireland and the suffrage was widened in New York, did the Irish begin to assume control of Tammany Hall.

About the only similarity between early and later Tammany was the criminal and corrupt practices of its leaders. When William Mooney, founder of Tammany, served as superintendent of the alms-house in New York from 1808 to 1809, he siphoned off so much money that food rations to the poor had to be cut substantially. As would happen later, the offending leader was not penalized by the society. After leaving the superintendent's office, Mooney was reelected Grand Sachem.

SOURCE: Daniel Boorstin, *The Americans: The Democratic Experience* (New York: Random House, 1973), pp. 256–57.

★ BOSS TWEED AND THE CARTOON ★

In 1875, Boss Tweed was in the Ludlow Street jail awaiting trial on charges that he had defrauded the government of millions of dollars. Thomas Nast, the cartoonist, who had spent years hounding Tweed and rousing public opinion against him, was triumphant. But Nast believed that Tweed would somehow find a way to escape all or most of the punishment he deserved.

"Tweed in Ludlow was allowed all sorts of liberties. He had the freedom of the city, and could drive out in the morning with a keeper for his coachman and a warden for his footman. In the evening he could dine at his Fifth Avenue home with a bailiff for his butler.

"It was at Tweed's home (Dec. 4) that he made his escape. The Deputy Sheriff had been invited to dine with him, and Tweed had requested that he might go up-stairs to see his wife. He did not return, and after hiding about New York for a time, fled to Cuba and eventually to Spain. That the great public offender in whose conviction [Nast] had been a chief instrument should have been allowed to escape was a humiliation to [the cartoonist]. . . .

"It had become known that Tweed was somewhere hiding in Spanish territory. As early as September 30, Nast cartooned him as a Tiger, appearing from a cave marked Spain. Now suddenly came a report—a cable—that one 'Twid' (Tweed) had been identified and captured at Vigo, Spain, on the charge of 'kidnapping two American children.'

"This seemed a curious statement; for whatever may have been the Boss's sins, he had not been given to child-stealing. Then came further news, and the mystery was explained. Tweed had been identified and arrested at Vigo through the cartoon 'Tweed-le-dee and Tilden-dum,' drawn by Thomas Nast. The 'street gamins'—to the Spanish officer, who did not read English—were two children being forcibly abducted by the big man of the stripes and club. The printing on the dead wall [behind Tweed] they judged to be the story of his crime. Perhaps they could even spell out the word 'REWARD.'

"Absurd as it all was, the identification was flawless. Tweed, on board the steamer *Franklin*, came back to America to die [in Ludlow, April 12, 1878]."

SOURCE: Albert Bigelow Paine, *Th. Nast: His Period and His Pictures* (New York: Macmillan, 1904), pp. 318, 336.

EDITOR'S NOTE: Great story. Would that it were true. The story circulated widely at the time and received confirmation in a Nast biography by Albert Bigelow Paine. Subsequently, scholars discovered that the story was untrue. Tweed actually had been identified by a low-level American counsel in Cuba a short time before he fled to Spain. His arrest in Spain was arranged by American authorities. "The outlandish legend of his being recognized from a Nast cartoon by some simple Spaniard," says Leo Hershkowitz, "is a complete fiction."

SOURCE: Leo Hershkowitz, *Tweed's New York: Another Look* (Garden City, N.Y.: Anchor Press, 1977).

The cartoon that captured Tweed. (Albert Bigelow Paine, *Th. Nast: His Period and His Pictures* [New York, Macmillan, 1904], p. 337.)

★ WON'T COMMIT HIMSELF ★

When Tammany Hall boss Charles F. Murphey didn't join a crowd in singing the national anthem at a Fourth of July celebration, a reporter became curious. An aide to the boss explained, "Perhaps he didn't want to commit himself."

SOURCE: William Riordan, *Plunkitt of Tammany Hall* (New York: Dutton, 1963), p. ix.

★ HONEST GRAFT ★

Tammany leader George Plunkitt:

"Everybody is talkin' these days about Tammany men growin' rich on graft, but nobody thinks of drawin' the distinction between honest graft and dishonest graft. There's all the difference in the world between the two. Yes, many of our men have grown rich in politics. I have myself. I've made a big fortune out of the game, and I'm gettin' richer every day, but I've not gone in for dishonest graft—blackmailin' gamblers, saloon-keepers, disorderly people, etc.—and neither has any of the men who have made big fortunes in politics.

"There's an honest graft, and I'm an example of how it works. I might sum up the whole thing by sayin': 'I seen my opportunities and I took 'em.'

"Just let me explain by examples. My party's in power in the city, and it's goin' to undertake a lot of public improvements. Well, I'm tipped off, say, that they're going to lay out a new park at a certain place.

"I see my opportunity and I take it. I go to that place and I buy up all the land I can in the neighborhood. Then the board of this or that makes its plan public, and there is a rush to get my land, which nobody cared particular for before.

"Ain't it perfectly honest to charge a good price and make a profit on my investment and foresight? Of course, it is. Well, that's honest graft."

Source: William Riordan, *Plunkitt of Tammany Hall* (McClure Phillips & Co., 1905), pp. 3–4.

★ A PRAYER FOR THE COUNTRY ★

"When Edward Everett Hale was Chaplain of the Senate, someone asked him, 'Do you pray for the Senators, Dr. Hale?' 'No, I look at the Senators and pray for the country,' he replied."

Mark Twain once remarked, "I think I can say, and say with pride, that we have legislatures that bring higher prices than anywhere in the world."

★ CARTOON ACCUSING BUSINESSMEN ★ OF HYPOCRISY

SIX DAYS WITH THE DEVIL AND ONE WITH GOD.
BUSINESS MAN TO CHRISTIANITY. "I am too Busy to see you Now. Wait till Sunday."

A cartoon from *Harper's Weekly*, 1869. (*Harper's Weekly*, July 3, 1869, p. 129.)

★ THE IMAGE-CONSCIOUS VICTORIANS ★

The chief characteristic of Victorianism was not moral virtuousness, but the appearance of moral virtuousness. The world did not have to be perfect, it only had to seem so. The necessity of maintaining a good image put quite a strain on people, especially those connected with morally dubious businesses. But people made do, as did the anonymous author of a guide to whorehouses in New York City, who resorted to an ingenious contrivance to give his book respectability.

Beginning his volume with a quotation from Shakespeare, he went on to explain that his little book was written to give the reader "an insight into the character and doings of people whose deeds are carefully screened from public view; when we describe their houses, and give their location, we supply the stranger with information of which he stands in need, we supply a void that otherwise must remain unfilled. Not that we imagine the reader will ever desire to visit these houses. Certainly not; he is, we do not doubt, a member of the Bible Society, a bright and shining light, like Newful Gardner or John Allen. But we point out the location of these places in order that the reader may know how to avoid them. . . . Our book will, therefore, be like a warning voice to the unwary—like a buoy attached to a sunken rock, which warns the unexperienced Mariner to sheer off, lest he should be wrecked on a dangerous and unknown coast." This explains why the book describes the houses in vivid detail and even reports on the beauty of the women and advises whether a letter of introduction is needed at any parlor.

SOURCE: *The Gentleman's Directory* (New York: privately published, 1870), *passim*.

★ OF SEVEN DEMOCRATS, NOW ONLY TWO ★

The news that Alferd E. Packer had murdered and eaten five hunting companions during a Colorado blizzard in 1873 was horrifying. Hardly anyone could think of a crime that was worse. But some people seemed concerned only because all of the victims happened to have been Democrats. One of these people was M. B. Gerry, judge at Packer's trial. When sentencing Packer, the judge flew into a rage. "Stand up, you man-eating son-of-a-bitch, and receive your sentence!" he began. "There were seven Democrats in Hinsdale County, but you, you voracious, man-eating son-of-a-bitch, you ate five of them. I sentence you to be hanged by the neck until you're dead, dead, dead, as a warning against reducing the Democratic population of the state."

SOURCE: *American Heritage*, October 1977, p. 112.

★ CUSTER AND THE INDIANS ★

General George Custer, commander of the ill-fated U.S. Seventh Cavalry at Little Big Horn, had great respect for the Plains Indians. He once confessed that he even enjoyed the heroic escapes of the Indians he pursued. Two years before Little Big Horn, Custer wrote: "If I were an Indian, I often think I would greatly prefer to cast my lot among those of my people [who] adhered to the free open plains, rather than submit to the confined limits of a reservation, there to be the recipient of the blessed benefits of civilization."

Custer's dead body was not mutilated on the Little Big Horn battlefield, an honor in Sioux warfare.

SOURCE: George Custer, *My Life on the Plains* (Norman, Okla.: University of Oklahoma Press, 1976), p. 22.

★ GENERAL BUTLER'S COURTROOM TACTICS ★

A reminiscence of Supreme Court Justice Oliver Wendell Holmes about Benjamin F. Butler, the Massachusetts lawyer, politician, and controversial Union general who declared escaped slaves to be contraband of war. Butler was a widely hated man and had a reputation for ruthlessness.

"General Butler was on his way to Boston to try a case before Judge Shaw. I met him on the train and asked him if I might look at the notes on the case. Butler acquiesced. To my astonishment I saw written on the top of the page, 'Insult the judge.' 'You see,' said Butler in answer to my question [about the line], 'I first get Judge Shaw's ill will by insulting him. Later in the case he will have decisions to make for or against me. As he is an exceedingly just man, and as I have insulted him, he will lean to my side, for fear of letting his personal feeling against me sway his decisions the opposite way.'"

SOURCE: Charles Shriner, ed., *Wit, Wisdom, and Foibles of the Great* (New York: Funk & Wagnalls, 1918), p. 93.

★ Butler's Loss, Republicans' Gain ★

When Butler, a Republican, failed to win reelection to Congress in 1874, many members of his own party were elated. Concerning the results of the election, in which Democrats swept to office, one Republican wired: "Butler defeated, everything else lost."

Source: John F. Kennedy, *Profiles in Courage* (1956; rpt. New York: Perennial Library, 1964), p. 116.

★ Uncle Sam at 100 ★

(*Leslie's Illustrated*, January 8, 1876, p. 296.)

★ Standing Up for Women's Rights ★

Reverend Henry Ward Beecher was chairing a meeting once when a woman stood up and began chanting a hymn of praise for women's rights. The woman went on and on, for more than a half hour. The audience grew visibly annoyed, but Beecher sat quietly and let her finish. When she finally sat down, the minister commented, "Nevertheless, brethren, I believe in women speaking in meeting."

SOURCE: Joseph B. Bishop, *Notes and Anecdotes of Many Years* (New York: Scribner's, 1925), p. 40.

★ BEECHER SELLS SOAP ★

Whenever he could, Beecher exploited his name to make money. The brother of moralist Harriet Beecher Stowe, he endorsed lingerie, Jay Gould's nefarious schemes, and even soap. One of his most profitable endorsements, which appeared in magazines throughout the country, was for Pears' Soap.

SOURCE: Burton Bledstein, *The Culture of Professionalism* (New York: Norton, 1976), p. 52.

HENRY WARD BEECHER wrote:

" If CLEANLINESS is next to GODLINESS, soap must be considered as a means of GRACE, and a clergyman who recommends MORAL things should be willing to recommend soap. I am told that my commendation of PEARS' Soap has opened for it a large sale in the UNITED STATES. I am willing to stand by every word in favor of it I ever uttered. A man must be fastidious indeed who is not satisfied with it."

PEARS' is the best, the most elegant and the most economical of all soaps for GENERAL TOILET PURPOSES. It is not only the most attractive, but the PUREST and CLEANEST. It is used and recommended by thousands of intelligent mothers throughout the civilized world, because, while serving as a detergent and cleanser, its emollient properties prevent the chafing and discomforts to which infants are so liable. It has been established in London 100 years as A COMPLEXION SOAP, has obtained 15 International Awards, and is now sold in every city in the world. It can be had of nearly all Druggists in the United States; but BE SURE THAT YOU GET THE GENUINE, as there are worthless imitations.

(*Life*, April 18, 1888, p. 238.)

★ PRESIDENT TILDEN? ★

When the residents of New York City awoke on the morning of November 8, 1876, they eagerly turned to their newspapers to discover who had won the presidential election held the previous day: Republican Rutherford Hayes or Democrat Samuel Tilden. Unfortunately, the contest was undecided, except so far as the New York *Tribune* was concerned, which announced in an unqualified headline that Tilden had been elected. The *Tribune*'s mistake may not have been as egregious as it seems. Three days after the election Hayes himself confided: "I think we are defeated in spite of recent good news. I am of the opinion that the Democrats have carried the country and elected Tilden."

SOURCE: Dee Brown, *The Year of the Century* (New York: Scribner's, 1966), p. 318.

★ RUTHERFORD B. HAYES HAS THE LAST LAUGH ★

When Rutherford B. Hayes became president in 1877, he immediately banished wine and liquor from the White House. Hayes was not a temperance fanatic, and until he became president had had no scruples about occasionally taking a drink. But when he moved into the White House, he decided to set a good example for the country and stop drinking.

Unfortunately, not everyone in America appreciated the President's action. Particularly perturbed were guests at White House dinners who wanted to drink but were not allowed to. They ridiculed Hayes and took to calling his wife "Lemonade Lucy," since the first lady refused to serve anything stronger than lemonade. A joke made the rounds in Washington that at Lucy's house the water flowed like wine.

Eventually guests at the White House found a way to get around Hayes's obnoxious prohibition. With the connivance of the stewards, they received punch made from oranges spiked with St. Croix rum. Somehow the stewards were able to smuggle the forbidden elixir into the White House without the knowledge of the President or his wife.

Or so all Washington thought. In his diary Hayes revealed that "the joke of the Roman punch oranges was not on us, but on the drinking people. My orders were to flavor them *rather strongly* with the same flavor that is found in Jamaica rum. This took! There was not a drop of spirits in them!"

SOURCE: Charles R. Williams, ed., *The Life of Rutherford B. Hayes* (Boston: Houghton Mifflin, 1914), II, 312–13n.

★ A WELCOMED DEATH ★

A story about an encounter between Senator George Hoar and William Evarts, secretary of state under Rutherford B. Hayes, as told by Henry Cabot Lodge:

"When he was in the Senate, my colleague in after years, Senator Hoar, had a bill about which he was very anxious, and which had been referred to Evarts for report. Months passed and no bill appeared. Meeting Mr. Evarts one day in the corridor, Mr. Hoar, who was his first cousin, said, 'By the way, Evarts, when you report that bill of mine, just notify my executors.' 'They will be gentlemen whom I shall be delighted to meet,' was the reply."

SOURCE: Henry Cabot Lodge, *Early Memories* (New York: Scribner's, 1913), p. 258.

★ EDISON FATHERS THE PHONOGRAPH ★

Thomas Alva Edison did not invent the phonograph to bring music to the masses. The great scientist was partially deaf and never cared for music. His reason for inventing the device had something to do with the telephone. Edison had helped Alexander Graham Bell develop the telephone by inventing the transmitter. But in early 1877 he began to worry that the telephone would be too expensive for the average American family. He immediately set to work to solve the problem.

Edison decided that if the telephone could not be brought to the people, the people would have to be brought to the telephone.

The invention would then be used in much the same manner as the old telegraph, with people bringing their messages to a central location. The only problem was that people could not leave their messages with the telephone as they had left them with the telegraph operator. With the telephone they were their own operators, communicating directly with the person they wanted to contact.

Edison thought he could easily solve this problem by inventing what he called a "telephone repeater." Within the year he had invented the phonograph. When he tested it, by shouting the verses of "Mary Had a Little Lamb" into the machine, it worked perfectly. "I was never so taken aback in all my life," he recalled later. "Everybody was astonished. I was always afraid of things that worked the first time." He received a patent for his invention in only two months because there was no evidence that anyone else had ever applied for a patent on a machine like Edison's. The creator of the light bulb had invented a device that no one else had ever even considered inventing.

The phonograph, as it turned out, was not used to help the telephone, which sometime afterward proved less expensive than Edison had feared. But the machine met with instant enthusiastic approval by the public. When it was exhibited in Boston, people came in droves to see it, paying more than $1,800 for tickets in one week alone.

Oddly, Edison did not foresee that his machine would be the basis of a major new industry—home entertainment. In 1878 he wrote in the *North American Review* that he believed the number-one use for the phonograph would be for "letter writing, and all kinds of dictation without the aid of a stenographer." He continued believing this for fifteen years. Finally, in 1894, he decided that the phonograph could be marketed widely as a music machine if it was sold for a low price. He managed to tag it at twenty dollars, but most people considered it too expensive—they preferred going to a public place where they could play a record for just a nickel. The time for the home phonograph had not yet arrived; Americans were still in the jukebox stage.

SOURCE: Daniel Boorstin, *The Americans: The Democratic Experience* (New York: Random House, 1973), pp. 379–80.

★ J. S. HARRISON GOES TO MEDICAL SCHOOL ★

John Scott Harrison is the only man in American history to be the son of one president and the father of another. So one would think that his corpse would have been well cared for. And it was, for a while. After dying peacefully in 1878, J. S. Harrison, who served two terms in Congress, received a proper funeral. A short time later, however, his body was discovered missing.

Throughout history medical colleges have rarely received enough cadavers for student experiments. By 1878 stealing recently deceased bodies from graveyards and selling them to medical schools had grown into a lucrative business. The nation was appalled, however, when a frantic search uncovered J. S. Harrison's body in the dissecting room of an Ohio medical college, patiently waiting its turn to help some medical student with his education. National shock soon turned to national wrath, forcing legislatures to enact stringent penalties against grave robbing. Harrison's body was returned to the morgue.

SOURCE: Sid Frank, *The Presidents* (Maplewood, N.J.: Hammond, 1975), p. 82.

★ MAXIMS OF MARK TWAIN ★

- "Man is the only animal that blushes. Or needs to."
- "You can straighten a worm, but the crook is in him and only waiting."
- "Well enough for old folks to rise early, because they have done so many mean things all their lives they can't sleep anyhow."
- "My books are water: those of the great geniuses are wine. Everybody drinks water."

★ MARK TWAIN'S SUPPRESSED MASTERPIECE ★

In 1879, Mark Twain delivered the following speech, "Some Thoughts on the Science of Onanism," before a Paris group called the Stomach Club. Until 1952 the speech remained in private hands, at which time a Chicago advertising executive came into possession of a copy and had it printed:

"My gifted predecessor has warned you against the 'social evil—adultery.' In his able paper he exhausted that subject; he left absolutely nothing to be said on it. But I will continue his good work in the cause of morality by cautioning you against that species of recreation called self-abuse, to which I perceive you are too much addicted.

"All great writers upon health and morals, both ancient and modern, have struggled with this stately subject; this shows its dignity and importance. Some of these writers have taken one side, some the other.

"Homer, in the second book of the *Iliad*, says with fine enthusiasm, 'Give me masturbation or give me death.' Caesar, in his *Commentaries*, says, 'To the lonely it is company; to the forsaken it is a friend; to the aged and to the impotent it is a benefactor; they that are penniless are yet rich, in that they still have this majestic diversion.' In another place this experienced observer has said, 'There are times when I prefer it to sodomy.'

"Robinson Crusoe says, 'I cannot describe what I owe to this gentle art.' Queen Elizabeth said, 'It is the bulwark of virginity.' Cetewayo, the Zulu hero, remarked, 'A jerk in the hand is worth two in the bush.' The immortal Franklin has said, 'Masturbation is the mother of invention.' He also said, 'Masturbation is the best policy.'

"Michelangelo and all the other old masters—Old Masters, I will remark, is an abbreviation, a contraction—have used similar language. Michelangelo said to Pope Julius II, 'Self-negation is noble, self-culture is beneficent, self-possession is manly, but to the truly grand and inspiring soul they are poor and tame compared to self-abuse.' Mr. Brown, here in one of his latest and most graceful poems, refers to it in an eloquent line which is destined to live to the end of time—'None know it but to love it. None name it but to praise.'

"Such as the utterances of the most illustrious of the masters of this renowned science, and apologists for it. The name of those who decry it and oppose it is legion; they have made strong arguments and uttered bitter speeches against it—but there is not room to repeat them here in much detail.

"Brigham Young, an expert of incontestable authority, said, 'As compared with the other thing, it is the difference between the lightning bug and lightning.' Solomon said, 'There is nothing to recommend it but its cheapness.' Galen said, 'It is shameful to degrade to such bestial uses that grand limb, that formidable member, which we votaries of science dub the "Major Maxillary"—when we dub it at all—when is seldom. It would be better to decapitate the Major than to use him so. It would be better to amputate the *os frontis* than to put it to such a use.' The great statistician Smith, in his report to Parliament, says, 'In my opinion, more children have been wasted in this way than in any other.'

"It cannot be denied that the high antiquity of this art entitles it to our respect; but at the same time I think that its harmlessness demands our condemnation.

"Mr. Darwin has grieved to feel obliged to give up his theory that the monkey was the connecting link between man and the lower animals. I think he was too hasty. The monkey is the only animal, except man, that practices this science; hence he is our brother; there is a bond of sympathy and relationship between us. Give this ingenious animal an audience of the proper kind, and he will straightway put aside his other affairs and take a whet; and you will see by his contortions and his ecstatic expression that he takes an intelligent and human interest in his performance.

"The signs of excessive indulgence in this destructive pastime are easily detectable. They are these: A disposition to eat, to drink, to smoke, to meet together convivially, to laugh, to joke, and to tell indelicate stories—and, mainly a yearning to paint pictures. The results of the habit are: loss of memory, loss of virility, loss of cheerfulness, loss of hopefulness, loss of character, and loss of progeny.

"Of all the various kinds of sexual intercourse, this has the least to recommend it. As an amusement it is too fleeting; as an occupation it is too wearing; as a public exhibition, there is no money in it. It is unsuited to the drawing room, and in the most cultured society it has long since been banished from the social board. It has at last, in our day of progress and improvement, been degraded to brotherhood with flatulence. Among the best-bred, these two arts are now indulged in only in private—though by consent of the whole company, when only males are present, it is still permissible, in good society, to remove the embargo upon the fundamental sigh.

"My illustrious predecessor has taught you that all forms of the 'social evil' are bad. I would teach you that some of these forms are more to be avoided than others. So, in concluding, I say, 'If you *must* gamble away your lives sexually, don't play a Lone Hand too much.' When you feel a revolutionary uprising in your system, get your Vendrôme Column down some other way—don't jerk it down."

SOURCE: Copy in New York Public Library.

★ IF IT'S ELECTRIC IT'S BETTER ★

Among the products of the late nineteenth century were Dobbins' Electric Soap (not to be confused, as an advertisement warned, with Magnetic Soap or Electro-Magic Soap), Philadelphia Electric Soap, and Dr. Scott's Electric Hair Brush. The products were not electrical, of course (what would electric soap be?), but the word "electric" was in their name because the word had a magic sound to it. Dr. Scott liked the word so much he even farsightedly sold an electric toothbrush—in 1880.

SOURCE: Edgar R. Jones, *Those Were the Good Old Days* (New York: Simon & Schuster, 1959), p. 25.

★ GARFIELD WRITES THE CLASSICS ★

James A. Garfield was a very talented and skillful man. Being ambidextrous, he could write with either his right or his left hand. Because of his classical education, the twentieth president was literate in both Greek and Latin. But the last president to be born in a log cabin had another ability. While writing Greek with one hand, Garfield could, at the same time, write Latin with his other hand.

SOURCE: Kevin McFarland, *Incredible!* (New York: Hart, 1976), p. 14.

★ JAMES GARFIELD'S EERIE TALK WITH THE SON ★ OF ABRAHAM LINCOLN

During his life James Garfield never voiced special concern about his own safety—except once.

The ex–Civil War soldier was a fatalist and believed, as he told a friend shortly after he was elected president, that assassination "can no more be guarded against than death by lightning; and it is not best to worry about either." But one day, four months into his term as president, Garfield suddenly and unexplainably became preoccupied with the possibility of death. He called in his secretary of war, Robert Todd Lincoln, to discuss the murder of the secretary's father.

Garfield, of course, knew all about the assassination of Abraham Lincoln. The irresponsible guard . . . the hole drilled in the wall behind the president's head . . . the escape of John Wilkes Booth over the edge of the balcony. Every American knew these facts. But Garfield wanted to know what the tragedy was like to someone on the inside, to someone who had known the touch of Lincoln's fingers and the slap of his hand.

With some difficulty the secretary of war recounted the story of the tragedy: the shock, the pain, the rush of events. Garfield asked a few questions and attempted to re-create the scene at Ford's Theater. He tried to imagine what it must have been like to be Booth, Robert Todd, or Lincoln himself. After a little more than an hour the secretary departed.

The meeting with Robert Todd Lincoln took place on June 30. On July 2, Charles Guiteau fired two shots at President Garfield and mortally wounded him—only two days after the President had expressed concern about death for the first time in his life.

SOURCE: Archie Robertson, "Murder Most Foul," *American Heritage*, August 1964, p. 91.

★ James Garfield Was Not Saved ★
by the Telephone

On July 2, 1881, President James A. Garfield was shot at the Washington railroad station. The first doctor on the scene gave him half an ounce of brandy and a dram of spirits of ammonia. Garfield promptly vomited.

D. W. Bliss, a leading Washington doctor, appeared and immediately tried to locate the path of the bullet in Garfield's body through the use of a heavy "Nelaton Probe." The instrument was introduced into the wound and turned slowly, probing for the point of least resistance, which, the doctor hoped, would be the bullet track. Unfortunately, the President was a muscular man and had been walking briskly when shot. Now he lay on his left side with muscles relaxed. When Bliss's probe suddenly slipped downward and forward three and a half inches into the President's body, it did not follow the course of the bullet. The instrument did, however, become stuck between the shattered fragments of Garfield's eleventh rib, and was removed only with a great deal of difficulty, causing the President terrific pain. Bliss next inserted his little finger into the wound, widening the hole in another unsuccessful probe for the bullet. When other expert doctors arrived from around the country, they, too, were unable to locate the bullet. A large black-and-blue spot did form, however, exactly where doctors assumed the bullet was and, coincidentally, where Dr. Bliss had originally probed.

Daily bulletins on the President's health worried the American public. One citizen, Alexander Graham Bell, inventor of the telephone, had a brainstorm. With one of his telephone receivers, he rigged up an electrical induction system with a primary and secondary coil, which, when brought into close proximity to a metal object, such as a bullet, created a slight disturbance in the balance of the circuit and sounded a faint hum in the telephone receiver. Because of Bell's worldwide fame, Garfield's doctors agreed to give the timely invention a try.

At the appointed hour, Bell arrived at the White House. The President was propped up in his bed and the equipment was put in place. While Bell stood behind the bed with the telephone receiver to his ear, an assistant slowly moved the coils around Garfield's abdomen and back. When the coils crossed the black-and-blue spot, Bell's face lighted up. He heard a hum. The experiment was repeated several times, once with Mrs. Garfield listening to the receiver, and each time a faint hum was detected as the coils crossed the black-and-blue area. The bullet was there, Bell confidently informed the President's doctors, but it was deeper than they had estimated. The doctors, however, judged that a major operation, when the exact location of the bullet was still somewhat uncertain, would be too hazardous.

But days later, when the President's temperature began to climb, the doctors reopened their previous incision and enlarged it outward and—as Bell had recommended—downward. Still they found no bullet. Garfield's condition wavered over a month longer. He finally died on September 17, 1881.

According to some medical historians, Garfield probably would not have died had he simply been left alone. What killed him, they say, was not the bullet, which became wrapped in a protective cyst, but infections caused by unsterile instruments and hands—infections that Garfield's body could not resist because the President had become weak after staying in bed for two months. Of course, Garfield's doctors, some of the best in the country, did not have the benefit of present-day knowledge.

And the bullet? In the autopsy it was located a full ten inches from where the doctors and Alexander Graham Bell had said it was. Bell's contraption had been no more accurate than a divining rod. Did Bell and Mrs. Garfield want to hear a hum so desperately that both simply imagined it? Or was it perhaps the sound of the President's metal-spring bed? It is impossible to say. Bell's invention, however, was not a total failure. Later, in a more elaborate form, it worked—for the army in the detection of land mines.

SOURCES: Rudolph Marx, *The Health of the Presidents* (New York: Putnam's, 1960), p. 242; *American Heritage Pictorial History of the Presidents of the United States* (New York: American Heritage Publishing Company, 1968), p. 530.

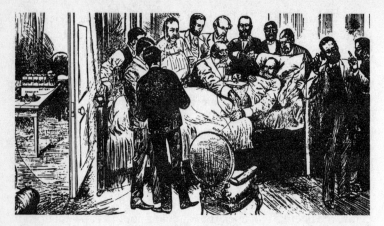

Bell trying to locate the bullet in Garfield's body. (*Leslie's Illustrated*, August 20, 1881, p. 421.)

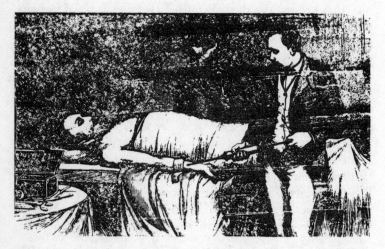

The corpse of President Garfield being embalmed. (*Leslie's Illustrated*, October 8, 1881, p. 85.)

★ THE NEAR DEATH OF A PRESIDENT ★

The best-kept secret of the Chester A. Arthur administration was that the president was terminally ill with Bright's disease and spent his last years in office knowing he could very well die before his term ended. Because of the unstable economic and political situation, Arthur reasoned that the public should not be told about his health problem. Already he labored under the burden of stepping into the White House upon the death of James Garfield. To announce that the new president might not live out his term would seriously damage his ability to govern. Instead, Arthur filled the White House with laughter and pretended nothing was wrong. The President knew that the more active he was the greater his chance of succumbing to the disease. Yet he even made a half-hearted attempt to gain his party's nomination for another term. Arthur survived his nearly four years in office, and did not die until a year and a half later, in 1886.

SOURCE: Thomas C. Reeves, *Gentleman Boss* (New York: Knopf, 1975), pp. 367, 374.

★ JESSE JAMES: THE ROBIN HOOD OF AMERICA ★

Folk legends surround the life of Old West outlaw Jesse James. Once, it has been told, while Jesse and his brother Frank were riding in the foothills of the Ozark Mountains with the Younger brothers, they stopped at a small, out-of-the-way cabin to ask for food. The sole occupant of the house was a poor, saddened woman whose husband had recently passed away. Overcoming any apprehension, the woman kindly agreed to throw some scraps together and feed the strangers. Once inside, however, Jesse sensed that something terrible was troubling the widow.

"Won't you tell us what's the matter?" he questioned earnestly.

After hesitating slightly the woman opened up. Her cabin, she wept, was mortgaged to the hilt. The banker, a heartless old skinflint, was expected that very day, and if she did not have the money he would foreclose. Sobbing, the woman confessed to her guests that she did not have a dime.

Jesse ate in silence, pondering the problem. Finally he asked, "How much do you owe this man?"

"Eight hundred dollars," came the answer.

"What does he look like and how will he be traveling?" Jesse asked.

The widow told him.

Unbeknownst to her, the James-Younger gang had been doing business in the Ozark area and Jesse had a considerable sum of money with him.

"It so happens," said Jesse, "I have that much money with me and I'm going to loan it to you."

The woman looked at him in startled disbelief, then began to weep again. "I'll work my fingers to the bone, but I don't know when I can pay you back," she cried.

"Don't you worry about that," Jesse reassured her. "I'll stop by sometime, then if you have it, you can pay me back."

"Now, you want to do this in a businesslike way," continued the outlaw, "so you ought to protect yourself. This gentleman here," he said, indicating his brother Frank, "will write out a receipt. Then you copy it in ink in your own handwriting. Before you pay over the money, you make the man sign the receipt. That's the proper way to conduct business. He'd make you do the same. And don't tell him anyone has been here. Now, will you do as I say?"

"Yes, sir," cried the astonished widow. "I think you are wonderful."

"I wouldn't say that," replied a humble Jesse. "I like to help deserving people when I can."

Later that afternoon, long after the James-Younger gang had departed, the skinflint banker called at the house of the widow. Disappointed that he would not be able to foreclose, he quickly took his eight hundred dollars, signed the receipt, and left. On his way home, however, about three miles from the widow's cabin, he was surprised by three mounted men. While one of the robbers held the banker's horse, another searched the man's belongings and, oddly enough, found eight hundred dollars. The outlaws then unbridled the banker's horse and sent it frantically galloping down the road, leaving him on foot. Their business finished, the thieves then rode away, eight hundred dollars richer. They were never captured or identified.

SOURCE: Homer Croy, *Jesse James Was My Neighbor* (New York: Duell, Sloan & Pearce, 1949), p. 100.

★ ONE CANNOT SHOOT WHITE MEN ★

The Old West range war between the farmers and the ranchers could be fairly unpredictable. Charley Coffee, an old Texas trail driver who had settled on a ranch in the western Nebraska–eastern Wyoming area, once told how everything there was pretty quiet until 1885, when "the Northwestern Railroad came poking in." Then the farmers, or "festive grangers," moved in, claiming, under the Homestead Act, 160-acre tracts of open land and surrounding them with barbed-wire fences. These farmer-pioneers created problems by squatting on the best watering holes in the territory and closing them off to the rancher and his cattle. The situation was delicate because, as Coffee noted, "It was not like the Indians for one couldn't shoot." So Coffee did the next best thing: he opened a bank. As he later explained to a friend, "The only way I could do to get even was to go into the banking business, so I am there." The homesteaders were generally poor folk, and Coffee's bank was more than willing to provide them with generous loans—at high interest. Sooner or later the struggling farmers would have trouble with their payments, and Coffee's bank would quickly foreclose on their mortgages. After a while the area's homesteader problem cleared up, as banker Coffee acquired under fee-simple ownership the same lands that rancher Coffee had earlier used as open range.

SOURCE: *Letters from Old Friends and Members of the Wyoming Stock Growers Association* (Cheyenne, Wyo.: S. A. Bristol Company, 1923), p. 28.

★ JUSTIFICATION FOR THE GHOST DANCE WAR ★

A month before he died, William R. Travers, the famous Indian fighter, discussed the army's implementation of U.S. policy with regard to the Indians. The last great Indian uprising, the Ghost Dance War, was in progress and a report had come about how General Nelson Miles's men had killed Sitting Bull.

"Been killing more Indians out west again, General," stated a friend who was reading the newspaper account of the Indian war.

"Yes," replied Travers, "the newspapers kill a good many Injuns. They kill more than the troops do. Why, if we killed half as many Injuns as the newspapers do, we'd be short of Injuns!"

"Is it right to kill these Injuns?"

"Dancing Injuns, ain't they? Ghost dancers?"

"Yes."

"Well, now," responded the general with mock gravity, "hasn't Sam Jones, and Moody, and the entire Methodist Church been trying to break up dancing for years? Of course they haven't succeeded. Now I'm glad that the strong arm of the government has at last united with the Church and taken hold of this dancing question. I hope General Miles will kill or convert every dancer west of the Mississippi, and then I hope the Secretary of War will call on General Howard to arrest the dancers, white or Injun, in the east—in New York and Philadelphia. I tell you dancing and chicken stealing must be stopped in this country."

SOURCE: Melville D. Landon, *Eli Perkins: Thirty Years of Wit* (New York: Cassell, 1891), p. 24.

★ THE HORSE WITH TWO LIVES ★

The real West and the dime-novel West rarely overlapped. But on December 18, 1890, at the death of Sitting Bull, the two worlds collided.

At the outbreak of the Ghost Dance War, the War Department ordered the arrest of Chief Sitting Bull. Although the Sioux chief was, by 1890, quite old and had lost much of his power within the tribe, the army still feared him as a great antiwhite leader. Forty-three Indian policemen, with the backing of two troops of the Eighth Cavalry just three miles away, were sent to make the arrest.

An hour before dawn the Indian police arrived at Sitting Bull's cabin. At first the old chief offered no resistance; but when a few policemen tried to speed things up by dressing him roughly, he became angry. A crowd of Sitting Bull supporters gathered. Almost ready to leave, the chief demanded that the Indian police saddle his horse.

Sitting Bull's was no ordinary horse, but an equine from the staged, showtime West. It had belonged to Buffalo Bill, and Sitting Bull had performed special tricks with it when the Indian had traveled with Buffalo Bill's Wild West Show. When Sitting Bull left the show to return to the Indian nation, Buffalo Bill, in friendship, gave the trick horse to him as a gesture of the showman's gratitude.

As the Indian police dragged and pushed Sitting Bull from his cabin, words were exchanged between the chief's supporters and his abductors. Suddenly, Sitting Bull announced that he was not going. Shots rang out, and with the first volley Sitting Bull was struck dead, one bullet entering from the front and another from behind. Both bullets were fired by the Indian police.

Oddly, when the shooting started, Sitting Bull's horse took the cue for his act in Buffalo Bill's Wild West Show. With bullets flying everywhere, Indian police and Sitting Bull partisans scurrying for cover, the horse began to perform his tricks. Right in the middle of the newly anointed battlefield, he sat down and raised one hoof. Terror-stricken, some of the Indian police thought Sitting Bull's freed spirit had entered his horse and made the animal do the act. The battle continued for thirty minutes. Fourteen people, from both sides, were killed.

Sitting Bull's horse, incredibly, was not injured, and an Indian policeman rode him to Fort Yates with news of the battle. Eventually, the chief's horse was returned to Buffalo Bill, who put him back to work in the Wild West Show. In 1893, at the Chicago Columbian Exposition, the Wild West Show's cavalry standard-bearer rode Sitting Bull's old horse.

SOURCE: Don Russell, *The Lives and Legends of Buffalo Bill* (Norman, Okla.: University of Oklahoma Press, 1960), p. 362.

★ The Glory That Was Harlem ★

The area of New York City north of Central Park that is called Harlem was not always an impoverished ghetto. In the middle of the nineteenth century Harlem was recognized throughout the country as the rural retreat of the rich. Cornelius Vanderbilt frequently visited Harlem and ran his horses there. Toward the end of the century Harlem became the center of one of the great real estate booms in American history. Brownstones went up by the dozens overnight, land prices soared, and three elevated railroads were built connecting Harlem to the center of Manhattan. Property became so valuable that descendants of the original owners of the land started a corporation in 1883 to prove their ownership of the commons and the marshes. Among the shareholders were General John C. Frémont and former Vice-President Schuyler Colfax. By 1893 the *Harlem Monthly Magazine* was boasting that in the near future the village would become "the center of fashion, wealth, culture, and intelligence."

But Harlem was actually on the precipice of ruin. Barely more than a decade later it was becoming the ghetto of the future. The main cause of decline was the construction of too many apartment houses. There simply were not enough tenants to occupy all the houses developers had built. When Philip Peyton, a Negro real estate salesman, offered to fill the apartments with black families, the landlords on some streets readily agreed. Of course, as blacks moved in whites moved out, and soon every street began being abandoned by whites. By the early 1920s, Harlem was virtually completely black.

Oddly enough, land values, which had been dropping in the 1910s, rose dramatically in the 1920s. Although whites had taken needed money out of the area, the entrance of blacks had eliminated the surplus of housing that had caused the original decline. Unfortunately, the flow of poor blacks into Harlem, swelled by great numbers of Negroes just then emigrating from the South, resulted in high demand for apartments and extravagant charges for rent. Inevitably, tenants began packing more people into their apartments than was healthy, landlords began to neglect their buildings, and unsanitary conditions developed. High rents had caused Harlem to become a congested ghetto.

SOURCES: Gilbert Osofsky, *Harlem: The Making of a Ghetto* (New York: Harper Torchbooks, 1966), *passim;* Daniel Boorstin, *The Americans: The Democratic Experience* (New York: Random House, 1973), pp. 294–95.

★ HOW ONE MAN FORCED ALL NEW YORKERS ★
TO STAY AT HOME

James Gordon Bennett, Gilded Age publisher of the New York *Herald,* once bragged to a group of friends that he could make the public do anything he wanted. To prove his claim, he boasted that the very next day he would make every New Yorker stay at home.

The following morning the city's streets were completely empty. Bennett had carried out his claim.

How had he done it? That morning the *Herald* had carried banner headlines announcing the escape of dangerous animals from the zoo. The headlines told of "Terrible Scenes of Mutilation" and "A Shocking Carnival of Death." Reportedly, animals were prowling everywhere, terrorizing the city. Finally, after several long, frightening hours, people realized they were the victims of a hoax. Slowly the streets filled up in the normal way and the city came to life.

SOURCE: *Scoundrels and Scalawags* (New York: Reader's Digest, 1968), p. 21.

★ CARNEGIE GIVES 16¢ ★

"Andrew Carnegie was once visited by a socialist who preached to him eloquently the injustice of one man possessing so much money. He preached a more equitable distribution of wealth. Carnegie cut the matter short by asking his secretary for a generalized statement of his many possessions and holdings, at the same time looking up the figures on world population in his almanac. He figured for a moment on his desk pad and then instructed his secretary, 'Give this gentleman 16¢. That's his share of my wealth.'"

SOURCE: Edmund Fuller, ed., *Thesaurus of Anecdotes* (Garden City, N.Y.: Garden City Publishing Company, 1943), p. 362.

★ The $50 Illness ★

Even the great are not above pecuniary concerns. Oliver Wendell Holmes returned the following message when asked to deliver a lecture: "I have at hand your kind invitation. However, I am far from being in good physical health. I am satisfied that if I were offered a $50 bill after my lecture, I would not have strength enough to refuse it."

SOURCE: Homer Croy, *What Grandpa Laughed At* (New York: Duell, Sloan & Pearce, 1948), p. 131.

★ Victorian Self-Abuse ★

Masturbation in the Victorian era was not acceptable behavior. It was a sin that robbed the spirit, decayed health, and led to insanity. In a report presented to the Massachusetts state legislature in 1848, the superintendent of the lunatic asylum at Worcester estimated that 32 percent of the asylum's patients were insane because of self-pollution. Similar figures were released by asylums around the country until the twentieth century.

The masturbator was easily identified. Plague spots, dark or blue spots under the eyes, were telltale evidence of a self-polluter. As one doctor put it, "they were the outward sign[s] of a morally bankrupt individual." But to the knowledgeable, many other clues indicated a possible self-abuser. Precocious physical development in an adolescent might very well have been triggered by masturbation. Children sliding on poles or trees were suspect. Tight-fitting clothes, dancing, or working long hours without proper exercise were also evidence of possible physical mistreatment.

The best preventive for masturbation was careful supervision by parents. Pure-minded parents were advised especially to watch their children during the vulnerable moments when a child goes to the bathroom, sleeps, or bathes. "Watch his motions as the child lies with covered head," suggested one specialist, "listen to his breathing. Is it quick, hurried, gasping, sighing? There is danger lurking there." Some boarding schools installed transoms above bathroom doors to enable the schoolmaster to keep an eye on suspected abusers.

Cures for the masturbator were more complicated. Sleeping in straitjackets or tying the practicer's hands either to the bedpost or to rings on the wall were usually effective. Special beds were designed which prevented a sleeper from turning over. A few ingenious Americans took out patents on the "genital cage," a metal truss that held the penis and scrotum with springs. To complement the effectiveness of the cage, it was recommended that people wear clothes which opened only from behind. But ingenuity did not stop there. One cage, patented in 1900 by a Mr. Joseph Lees of Pennsylvania, featured an electrical alarm that was triggered in the unfortunate event of an erection.

What today would be considered barbaric cures for the self-polluter were recommended and practiced by Victorians. According to some doctors, leeches placed around the sexual parts helped remove "congestion." Bloodletting also quelled a masturbator's appetite for sin. Other cures were more extreme. Some specialists prevented masturbation by perforating the foreskin of an uncircumcised penis and inserting a ring through it. Drugs were applied to the sexual parts to make a would-be abuser too sore to masturbate. After a strong dose of red iron or Spanish fly was applied to the genitals, masturbation was no longer a great temptation.

But some doctors resorted to even more radical measures. Hypothesizing that one-half of the females who masturbated did so because of an irritation of the clitoris, one Chicago doctor recommended clitoral circumcision as a sure cure for female self-abuse. An Ohio doctor told in 1896 of the success he had experienced with an operation on habitual male abusers in which he removed from half an inch to an inch of the dorsal nerves in the penis. The operation rendered patients sexually impotent for approximately a year and a half, enough time to permit, in the doctor's words, the "restoration of the physical and mental health."

Occasionally, nineteenth-century specialists prescribed marriage as the cure for male and female masturbation. Individual self-abuse, it seems, was practically eliminated by this surest of all Victorian remedies.

SOURCE: John S. Haller Jr., and Robin M. Haller, *The Physician and Sexuality in Victorian America* (Urbana: University of Illinois Press, 1974), p. 223.

★ WOMEN ARE DUMBER ★

In the late 1900s many physicians regarded increased female education as a primary factor in a general decline of female health. A woman's brain was simply not capable of assimilating a great deal of academic instruction. Education past high school, many specialists warned, was both physically and mentally destructive to the female. A study published in the *Medical Record* in 1892 illustrated the problem. Of 187 high school girls diagnosed, 137 constantly complained of headaches—clear evidence, concluded the report, of female inability to deal with the complexities of a rigorous academic program.

The rising number of neurotic girls, young women afflicted with emotional or psychic disorders, was directly linked to increased female education, many doctors believed. The affliction was often characterized by fatigue, depression, feelings of inadequacy, and other physical and mental ailments. Young women, commented one physician, "whose mental powers are overtaxed before their brains are sufficiently developed," were the most likely individuals to break down in nervous exhaustion.

Even though a high percentage of specialists believed the female brain was simply not made to perform intellectually, a woman's natural constitution did make her much less susceptible to many physical abnormalities that commonly afflicted men. Senility, loss of sight or hearing, and a host of other ailments were primarily associated with men. Doctors warned, however, that female efforts to imitate the male would destroy a woman's inherited immunity to certain maladies. Already, doctors reported, male afflictions such as paralysis, insanity, alcoholism, and crime, which were caused by overwork or prolonged worry, were on a frightful upswing among the gentler sex.

SOURCE: John S. Haller Jr., and Robin M. Haller, *The Physician and Sexuality in Victorian America* (Urbana: University of Illinois Press, 1974), p. 37.

★ AN OLD STORY: WARNINGS AGAINST SMOKING ★

An advertisement from 1892. (Edgar R. Jones, *Those Were the Good Old Days* [New York: Simon & Schuster, 1959], p. 56.)

★ WEIRD WHITE HOUSE RULES ★

A few of the rules and maxims appearing in the 1887 edition of the official White House book of etiquette:

- "A gentleman should not bow from a window to a lady, but if a lady recognizes him from a window, he should return the salutation. It is best, however, for a lady to avoid such recognitions. It is not in the best taste for her to sit sufficiently near her windows to recognize and be recognized by those passing on the street."
- "Cleanliness is the outward sign of inward purity. It is not to be supposed that a lady washes to become clean but simply to remain clean."

- For men: "Do not indulge in long hair, thinking it gives you an artistic look. Except in painters and poets, flowing locks are a ridiculous affectation."

SOURCE: Janet Halliday Ervin, *The White House Cookbook* (1887; rpt. Chicago: Follett Publishing Company, 1964), *passim*.

★ CLEVELAND THE EXECUTIONER ★

Before he became president, Grover Cleveland served as sheriff of Erie County, New York. Twice during his tenure the future president was called upon to hang convicted criminals. Cleveland, who would be president only fourteen years after accepting the job as sheriff, actually placed the noose around the necks of the convicted men, tightened the rope, and sprang the trapdoor.

SOURCE: Joseph Kane, *Facts about the Presidents*, 2d ed. (New York: H. W. Wilson, 1968), p. 151.

★ THE PRESIDENT'S SECRET OPERATION ★

On the evening of June 30, 1893, the President of the United States quietly slipped on board a yacht anchored at Pier A on the East River in New York City. Mystery surrounded his appearance. The crew had been told the President would be having two teeth pulled the next day, but the extreme measures taken to keep his presence a secret seemed strange.

Some time earlier Grover Cleveland had complained about a rough spot on the roof of his mouth. His doctors had examined it and found it to be cancerous. They had advised that he have an operation to have it removed. Part of his jaw would have to be cut out, but within a few weeks Cleveland would be fully recuperated. There would be no sign of an operation, since the surgery would all be done from the inside of his mouth.

The operation took place on July 1 and went smoothly. The upper part of the jaw was removed, and within two days Cleveland was up and about. But he could not speak well, and after the cotton that had been placed in the excavated area had been taken out, his speech became completely unintelligible. He recovered quickly, however, though a second operation had to be performed on a remaining area of cancer.

The public was not told about the operation, since Cleveland believed news of his illness might lead to uncertainty and worsen the financial crisis that was then developing. Congress was about to begin debate on the repeal of the Sherman Silver Act and nerves on Wall Street were tense. On August 29, however, an amazingly accurate report of the operation in all its details appeared in the Philadelphia *Public Ledger*. Apparently one of the doctors had broken his oath and revealed everything.

But by then Cleveland looked just as he always had. An artificial jaw made of vulcanized rubber preserved his familiar jowly expression and made the story seem ridiculous. His doctors condemned the report, and friends asserted that the President never looked better in his life. Newspapers around the country published attacks on the *Ledger*'s story, and in a few weeks it was virtually forgotten.

For almost twenty-five years the public heard nothing more about the operation. Then, in 1917, nine years after Cleveland's death, one of his doctors told all, in an article published in the *Saturday Evening Post*.

SOURCE: Rudolph Marx, *Health of the Presidents* (New York: Putnam, 1960), pp. 253–62.

★ MARK HANNA'S PHILOSOPHY OF LIFE ★

When a young Republican prosecutor from Ohio began a suit in 1890 to void the charter of the Standard Oil Company, Mark Hanna, the Republican party's Mr. Moneybags, became irate. To the young prosecutor Hanna condescendingly wrote: "You have been in politics long enough to know that no man in public office owes the public anything."

SOURCE: Matthew Josephson, *The Robber Barons* (New York: Harcourt, Brace, 1934), p. 353.

★ NEW YORK PRIEST BATTLES TAMMANY HALL ★

On February 14, 1892, a Sunday, Charles Parkhurst, pastor at a New York City church, delivered a sermon against venality in municipal government. Without naming names, Parkhurst insinuated that virtually every bigwig in the city was corrupt. The politicians? They were a "damnable pack of administrative bloodhounds . . . fattening themselves on the ethical flesh and blood of our citizenship." The Tammany bosses? "They are a lying, perjured, rum-soaked, and libidinous lot" in league with the Devil himself.

It was the kind of sermon seldom heard in America, at least in the cities, during the last third of the nineteenth century. A sermon full of fire and brimstone, burning exclamation points, and searing indictments. It could not be ignored.

And it wasn't. Within a few days a grand jury was impaneled; nine days later Parkhurst was called to testify; by the fifteenth day the jury issued a report.

Their conclusion? The city's problems could all be blamed on one person: Charles H. Parkhurst. Because of him the people of the city were seething with indignation. He had tempted them to distrust their leaders and had made charges he could not prove. When the pastor had faced the grand jury, he had brought with him a balloon filled with nothing but hot air. The New York *Sun* suggested that Parkhurst be sent to prison.

The fact was that Parkhurst had not been ready to talk when called before the grand jury. All he knew was what everyone knew: the city was a cesspool of corruption. Was the district attorney in on it? Was the mayor? How about the justices of the court? Which ones? Parkhurst couldn't say. And so nothing changed.

SOURCE: E. M. Werner, *It Happened in New York* (New York: Coward-McCann, 1957), pp. 36–116.

★ MURDER AT THE COLUMBIAN EXPOSITION ★

Herman Webster Mudgett, alias H. H. Holmes, killed more young women than anyone else in American history. In 1892 Holmes purchased a vacant lot across the street from his Chicago drugstore. On the site he designed and constructed a three-story home and office building that would later be known as "Murder Castle." Firing work crews as they finished each individual part, Holmes was able to keep the master plan of the house a secret. When finished, the house contained secret rooms, concealed stairways, trapdoors, false walls and ceilings, doors that opened to solid brick walls, an elevator shaft with no elevator, an elevator with no shaft, and a hidden chute descending from the third floor to the basement. Holmes, a former medical student at the University of Michigan, equipped the basement with a mammoth dissecting table, a stone crematory, and large vats of quicklime and acid.

Once the house was built, Holmes began looking for victims. At employment agencies throughout the city he advertised for secretaries. "He liked nice, green girls fresh from business college," one account later reported. It was the summer of 1893, and the Chicago Columbian Exposition was rolling into full swing. The bustling, excited city provided the perfect environment for Holmes's criminal activities. To each new secretary Holmes would promise marriage and love forevermore. In return, the beguiled female would gladly sign over all insurance policies and savings accounts, and write out a will naming Holmes as beneficiary. The "budding romance" would culminate in Holmes's third-floor bedroom, where the loving couple would spend the night. The next morning, however, Holmes would awaken early and, with chloroform from his laboratory, deepen the sleep of his lover. Carefully removing the girl to the elevator shaft, he would then wait for her to wake up. Next he would gleefully watch as lethal gas was pumped into the chamber, causing the betrayed girl to claw and gasp for help. When it was all over, he would throw the dead body down the secret chute to the basement.

After a year or so Holmes grew tired of killing women and began working at other things. The hapless girls simply were not bringing in enough money. "Lord knows I've worked hard," he once admitted to an accomplice, "but the damnable place has cost me $50,000 to operate. I'm going broke in this business."

In 1895, Holmes was arrested in Philadelphia in an insurance swindle involving murder. Searching for clues in that investigation, detectives opened "Murder Castle" and discovered over two hundred corpses, in varying stages of decay.

H. H. Holmes confessed all, but his memoirs, which he was dictating to the newspapers, were cut short when he was only up to his twenty-seventh victim. At that point, May 7, 1896, he was hanged for the crime of murder.

SOURCE: Jay Robert Nash, *Bloodletters and Badmen* (New York: M. Evans and Company, 1973), p. 382.

★ HE MADE A CRIME WAVE ★

In his autobiography Lincoln Steffens, the great nineteenth-century reporter, made the following statement: "I enjoy crime waves. I made one once."

It happened this way. One summer day in New York City in the 1890s, Steffens was wiling away the afternoon sitting in the cool basement of police headquarters listening to stories told by criminals, cops, and reporters. Suddenly a bored detective interrupted and said that he had a really interesting story to tell.

A prominent family living on Fortieth Street and Madison had gone away for the summer and entrusted their house to a caretaker, named Billy Bones. Unfortunately, Billy was not very trustworthy and had connived to become caretaker in order to rob the family blind. Accordingly, shortly after the family left the city, he and a friend, Mr. Busy-Bee, who arrived at the house with a wagon, began clearing things out. There was one small problem, however: two policemen standing on a nearby corner. "Oh, well," Billy said when Busy pointed out the cops, "they're not Chicago bulls; they're only New Yorkers. If they come up, we'll ask 'em to help us. See?"

"All right, Billy," Busy said. "We'll try it; I don't want to hire the wagon twice for nothin'. Let's get some heavy things down on the sidewalk so as to give them something to do."

Sure enough, soon one of the cops walked up and questioned the two thieves about the clutter on the sidewalk. Billy identified himself as the caretaker and explained that they were trying to get the furnishings out of the house as quickly as possible. Would the policeman, he asked, be so good as to help them load up the things?

The man in blue hemmed and hawed and said he had not joined the force to be someone's moving man. "Ah, come on," Billy implored, "be a sport an' give me a boost with this trunk." The cop finally relented. He helped with the trunk and a few other items, and then loaded up a parlor clock all by himself. Soon everything was in the wagon and the two men bade farewell to their well-meaning public servant.

Ordinarily Steffens never bothered with crime news, but this story concerned an important family connected with Wall Street. So he wrote up the account for his paper, the *Post*, though he good-naturedly omitted the joke played on the cops.

The story immediately created a sensation, not so much on the streets as in the offices of the city's other papers. At this time the practice of the city's papers had been to obtain all their news about crime from court records. Thus, all the papers always carried the same stories about criminal activities. Suddenly, however, one paper had got a scoop. This led editors at the other newspapers to demand that their reporters also come up with stories which no one else had.

In no time, of course, every paper was outscooping the other, reporting all types of crimes that before would have gone unnoticed. Particularly adept at finding new crimes was Jacob Riis, who had a source in the Police Department that supplied him with virtually every report of illegal activity anywhere in the city. Every day Riis would report on three or four new crimes, while Steffens would tell about one or two more that he had discovered, with other newsmen contributing a few stories of their own.

After a few days the people of New York began to think, naturally, that their city was suffering from a major crime wave. Were the criminals taking over the city? Respect for the newly appointed police board declined dramatically. Its head, Theodore Roosevelt, a friend of both Steffens and Riis, began worrying that his attempts to clean up the city had somehow started it. Under the old, corrupt system police had an informal agreement with the city's criminals not to harass them too much if they kept criminals from other cities out of the metropolis. T.R. had broken this long-standing agreement, demanding that anyone found breaking the law be arrested. Now he wondered if he had been right.

At a secret meeting of the police board, the police commissioner told T.R. that Roosevelt could stop the crime wave anytime he wanted.

"I! How?" Roosevelt stuttered.

"Call off your friends Riis and Steffens. They started it, and—they're sick of it. They'll be glad to quit if you'll ask them to."

"I don't understand," T.R. replied.

The commissioner explained that he had checked the record of arrests and found that city crime had not increased one bit. The only thing that had increased was newspaper reports of crime. After learning this, the commissioner had talked with some newspaper friends of his and asked about the increase in crime news. They had informed him that it was all the fault of Riis and Steffens, who had started competing for the most stories about crime.

T.R. ended the meeting and called Riis and Steffens to his office. "What's this I hear?" he shouted wildly. "You two and this crime wave? Getting us into trouble? You? I'd never have believed it. You?"

Riis and Steffens explained what had happened, and T.R. told them to stop competing for stories. Neither reporter liked rushing around to get a scoop, so they both quickly agreed to Roosevelt's suggestion. And that effectively ended New York City's great crime wave.

SOURCE: Lincoln Steffens, *The Autobiography of Lincoln Steffens* (New York: Literary Guild, 1931), pp. 285–91.

★ IDAHO ELECTS A SENATOR? ★

The Seventeenth Amendment, ratified in 1913, provided that U.S. senators be elected by a statewide popular vote. Before that time senators were selected by the state legislatures. This arrangement led occasionally to curious politics, as in the case of the 1896 election of Henry Heitfeld to the U.S. Senate from the state of Idaho.

The 1896 Idaho legislature consisted of Republicans, Democrats, and Populists. Normally, the Democrats and Populists formed a coalition, but this particular year there were so many aspirants for the Senate seat that no one could build a strong base of support. Heitfeld was a Populist who at times received a number of votes himself. But Heitfeld personally seemed to be a rather unambitious candidate, always voting for another man. One day there was a particularly large vote cast for Heitfeld, but no one dreamed he might be close to being elected. Heitfeld sat quietly at his desk, voted for someone else, and made his own personal tally of the votes. When the roll call ended, however, he stood up and addressed the chair: "Mr. President, I desire to change my vote. I vote for Heitfeld."

A loud roar of laughter filled the chamber, and Heitfeld immediately became the butt of a great deal of chafing from his fellow legislators. Meanwhile, the clerks compiled the official tally, and before anyone, except Heitfeld, realized what was happening, the presiding officer declared: "Henry Heitfeld having received a majority of all the votes is hereby declared elected United States Senator for the term of six years."

Heitfeld, of course, had tallied his votes and seen that chance had thrown him to within one vote of the Senate chair. So he changed his vote and earned a trip to Washington, D.C.

SOURCE: Arthur Wallace Dunn, *From Harrison to Harding* (New York: Putnam, 1922), p. 219.

The Full Dinner Pail, the Bull Moose, and the Great War

"I want to be a Bull Moose
And with the Bull Moose stand
With Antlers on my forehead
And a Big Stick in my hand."
                    —Bull Moose Party Jingle

★ Scrapbook of the Times ★

- When it became clear that William McKinley had been elected president over William Jennings Bryan in 1896, campaign manager Mark Hanna wired McKinley: "God's in his Heaven—all's right with the world."
- In 1897 the federal government recalled $26 million worth of one-hundred-dollar bills when a counterfeit hundred-dollar bill appeared that was so accurate it almost could not be distinguished from the real thing. This was the only time in history that fake money was so well designed that legitimate currency had to be withdrawn.
- Whenever his wife suffered an epileptic seizure in public, there was always one thing William McKinley would do: throw a napkin over her face.
- McKinley did not want to be the first president to leave the boundaries of the United States during his term. So when he took a walk in 1901 on the bridge connecting the United States and Canada at Niagara Falls, he was careful not to go more than halfway.
- Until 1900 the state of Rhode Island had two capitals, at Providence and Newport.

- During the Spanish-American War, Americans were portrayed in Spanish cartoons as pigs—because the United States was a big exporter of pork to Cuba.
- The safety razor, the subway, the long-distance telephone, and the movies all made their appearance during the Progressive Era.
- During the first five years of the twentieth century a Negro was lynched almost every other day.
- Despite the primitiveness of the practice of bleeding, it was still resorted to as late as 1905, when the Sears catalogue carried this advertisement: "Spring Bleeding Lance. The only practicable, safe and convenient instrument for bleeding on the market. Used almost exclusively by old school physicians for the purpose."
- One C. K. G. Billings held a dinner party at which everyone dined while sitting on a horse. The party cost $250 per person, much of the expense going for custom-made trays used to hold the food on each animal.
- When Douglas MacArthur left home to attend West Point, his mother went with him—to keep an eye on the future soldier. To make sure he studied, she took an apartment that had a perfect view of his dormitory room.
- J. P. Morgan gave this answer when asked about the cost of maintaining his yacht, the *Corsair:* "Nobody who has to ask what a yacht costs has any business owning one."
- In 1906, Maxim Gorky and wife were ousted from a New York apartment building when they could not persuade the proprietor that they were married.
- Upon entering the Senate in 1906, the insurgent Republican Robert LaFollette was appointed by the Old Guard to a committee that had never met: the Committee to Investigate the Condition of the Potomac River Front.
- Stopping in Nashville at the Hermitage, the home of Andrew Jackson, after a bear-hunting trip in 1907, Theodore Roosevelt coined a famous phrase when, after drinking a cup of Maxwell House coffee, he remarked—according to the General Foods Corporation—"Delighted—this coffee is good to the last drop."

- When the Great White Fleet, which T.R. sent around the world in 1907, returned to America in 1909, it was immediately painted gray.
- One of Teddy Roosevelt's sons once remarked, "When father goes to a wedding, he wants to be the bride; when he goes to a funeral, he wants to be the corpse."
- In 1908 the New York Giants lost the pennant to the Chicago Cubs because player Fred Merkle, in rounding the bases, missed second.
- The Sears catalogue of 1909 carried an advertisement for a horseless buggy that was "guaranteed to go 100 miles in 24 hours if good care is taken of it."
- When King Edward VII of England died in 1910, the New York Stock Exchange closed for the day.
- In 1910 a down-and-out young Italian named Mussolini almost emigrated to the United States.
- After his first night with his second wife, Woodrow Wilson broke into a dance and sang out the lines to the song, "O What a Beautiful Doll."
- America's first self-service grocery store, the Piggly Wiggly, opened in 1916. The store was so organized that customers had to go up and down every aisle before reaching the checkout counter.
- During World War I, Wilson raised $100,000 for the Red Cross by selling the wool of White House sheep. The sheep had been purchased at the beginning of the war to replace the gardeners who were drafted by the army.
- After listening to a Senate debate on what the needs of the country were, Vice President Thomas Marshall coined the memorable expression: "What this country needs is a really good five-cent cigar." (In the 1930s, Franklin P. Adams gave the phrase a twist when he remarked: "What this country needs is a good five-cent nickel.")
- In the fall of 1918 the Justice Department arrested 75,000 civilians in two days on suspicion of draft dodging. It turned out that only 3 percent of the men were illegally out of uniform.

- John D. Rockefeller: "God gave me money."
- On September 1, 1918, the secretary of war ordered the baseball season cut short on account of World War I.
- World War I ended at precisely eleven o'clock on the eleventh day of the eleventh month of the year 1918.
- When a militaristic congressman asked a woman testifying for peace at a congressional hearing, "Who won the World War?" colleague Maury Maverick shouted out the famous line: "Who won the San Francisco earthquake?"
- When Georges Clemenceau was told about Woodrow Wilson's Fourteen Points, he exclaimed: *"Le bon Dieu n'avait que dix!"* (The good Lord had only ten!)
- At the White House, through the early years of the twentieth century, a man calling on the president had to present two visiting cards, one for the president and one for the first lady. A woman visitor had only to present one card, since she was presumably calling only on the president's wife.

★ HEINZ'S 57 VARIETIES ★

In 1896, Henry John Heinz was riding on an elevated railway in New York City when he saw an advertisement for twenty-one varieties of shoes. The sign gave him an idea. He would advertise his own company's products with a number—any number, so long as it was catchy. Finally he decided upon fifty-seven, which he believed people would remember. The number itself was meaningless, of course. Even in 1896 the Heinz company sold more than fifty-seven varieties.

SOURCE: Alex Groner, *The American Heritage History of American Business and Industry* (New York: American Heritage Publishing Company, 1972), p. 255.

★ THE CURE FOR AMERICAN ILLS ★

In 1898 the Bayer pharmaceutical firm introduced into drugstores across the country its new medication for bad coughs—heroin. It

An advertisement from 1923. (Edgar R. Jones, *Those Were the Good Old Days* [New York: Simon & Schuster, 1959], p. 336.)

was not the first "hard" drug to be pushed on the American market. Lack of government regulation, easy accessibility, and medical ignorance made drug abuse a major social problem in the late 1800s. Common pharmaceutical products included many dangerous substances. Cocaine tablets for throat and nerves; baby syrups spiked with morphine; miscarriage-producing Portuguese Female Pills, "a great and sure remedy for married ladies"—were all readily available, either by mail or at the local drugstore. The base of most liquid medications was alcohol. Consumption was so great that experts have estimated that nineteenth-century Americans imbibed more spirits from patent medicines than from bottled liquor. Not until 1909, with the enactment of the Narcotics Drug Act, did the government begin to regulate the quality of medications.

Drug advertisements were as plentiful as the products they promoted. The pages of *Life*, *Harper's*, the Sears catalogue, and the New York *Times* were filled with a virtual onslaught of drug-touting notices. According to Congressman William Everett of Massachusetts, one church congregation found the advertising barrage to be especially trying. New hymnals were needed, but the church had little money to buy them. To economize, the congregation contracted with a patent-medicine manufacturer who agreed to defray a large percentage of the hymnal cost in return for advertising space in the new books. The songbooks arrived on December 24. On Christmas Day the churchgoers filled the sanctuary only to find in their new hymnals:

> *Hark! The herald angels sing*
> *Beechan's pills are just the thing.*
> *Peace on earth and mercy mild*
> *Two for man and one for child.*

SOURCES: Otto L. Bettmann, *The Good Old Days: They Were Terrible* (New York: Random House, 1974), p. 152; Edward Boykin, ed., *The Wit and Wisdom of Congress* (New York: Funk & Wagnalls, 1961), p. 376.

Advertisement from an 1880s edition of *Life* magazine. (*Life*, January 21, 1886, p. 56.)

★ ANYTHING YOU WANT FROM ★
MONTGOMERY WARD

In the late nineteenth century Montgomery Ward's catalogue offered an astonishing variety of merchandise. Many people believed they could order anything they wanted from Ward. Even a wife. One man wrote: "As you advertise everything for sale that a person wants I thought I would write you, as I am in need of a wife, and see what you could do for me." Another man, more demanding, wrote: "Please send me a good wife. She must be a good housekeeper and able to do all household duty. She must be 5 feet 6 inches in height. Weight 150 lbs. Black hair and brown eyes, either fair or dark. I am 45 years old, six feet, am considered a good-looking man. I have black hair and blue eyes. I own quite a lot of stock and land. I am tired of living a bachelor life and wish to lead a better life and more favorable. Please write and let me know what you can do for me."

SOURCE: Daniel Boorstin, *The Americans: The Democratic Experience* (New York: Random House, 1973), pp. 123–24.

★ HEARST'S WAR ★

William Randolph Hearst was looking around for a way to boost sales of his newspaper when he discovered Cuba. A war in Cuba would excite the public and would, in turn, create a great demand for newspapers. Late editions . . . extras . . . irresistible headlines. So Hearst sent the great Western painter, Frederic Remington, to Cuba to get pictures of "a gallant revolution." Remington went but found that there was no revolution. "Everything is quiet," he wired. "There is no trouble here. There will be no war. I wish to return." "Please remain," Hearst cabled back. "You furnish the pictures and I'll furnish the war."

SOURCE: Curtis MacDougall, *Hoaxes* (New York: Macmillan, 1940), p. 244.

★ YANKEES DEFEATED IN ★ SPANISH-AMERICAN WAR

General Joseph Wheeler, who had been a Confederate cavalry officer in the Civil War, served admirably in the Spanish-American War, but he had a hard time remembering who the enemy was at the Battle of Santiago. He went into the battle though he was very sick and had to be carried in an ambulance. When the battle seemed to be going badly, he bravely left the ambulance, dramatically leaped on a horse, and led a charge. The charge was succeeding when Wheeler, slipping back into his youth, shouted exultantly to his men, "The Yankees are running; they are leaving their guns!" "Oh, damn it," he corrected himself when he remembered where he was and what uniform he was wearing, "I didn't mean the Yankees, I meant the Spaniards!"

SOURCE: Champ Clark, *My Quarter Century of American Politics* (New York: Harper & Brothers, 1920), II, 45.

★ T.R. ON KETTLE HILL ★

If there is one thing that everybody knows about the Spanish-American War and the Battle of Santiago, it is that Teddy Roosevelt led the Rough Riders in a charge up San Juan Hill. Actually, T.R. and his invincible Rough Riders charged up Kettle Hill, a smaller mound in front of and off to the right of San Juan Hill. From Kettle Hill the Rough Riders fired on the Spaniards, and the regular infantry started up San Juan Hill. T.R. later wrote in his story about the Rough Riders that Kettle Hill afforded him and his troops "a splendid view of the charge on the San Juan Blockhouse." By the time Roosevelt had descended Kettle Hill and followed the rest of the American army force up San Juan Hill, the Spaniards had long since fled.

SOURCE: Charles H. Brown, *The Correspondent's War* (New York: Scribner's, 1967), p. 358n.

★ "ALONE IN CUBIA" BY T.R. ★

At the turn of the century Finley Peter Dunne was one of America's leading satirists. One of his favorite targets was Theodore Roosevelt. In one article he has "Mr. Dooley," the man whom he speaks through, tell "Mr. Hinnissy" about a book on the Spanish-American War written ("Mr. Dooley" says) by T.R. "Mr. Dooley" has the Rough Rider "tell th' story in his own wurruds":

""""We had no sooner landed in Cubia than it become nicessry f'r me to take command iv th' ar-rmy which I did at wanst. A number of days was spint be me in reconnoitring, attinded on'y be me brave an' fluent body guard, Richard Harding Davis. I discovered that th' inimy was heavily inthrenched on th' top iv San Joon hill immejiately iv front iv me. At this time it become apparent that I was handicapped be th' prisence iv th' ar-rmy," he says. "Wan day whin I was about to charge a block house sturdily definded be an ar-rmy corps undher Gin'ral Tamale, th' brave Castile that I afterwards killed with a small ink-eraser that I always carry, I r-ran into th' entire military force iv th' United States lying on its stomach. 'If yet won't fight,' says I, 'let me go through,' I says. 'Who ar-re ye?' says they. 'Colonel Rosenfelt,' says I. 'Oh, excuse me,' says the gin'ral in command (if me mimry serves me thrue it was Miles) r-risin' to his knees an' salutin'. This showed me 'twud be impossible f'r to carry th' war to a successful con-clusion unless I was free, so I sint th' ar-rmy home an' attackted San Joon hill. Ar-rmed on'y with a small thirty-two which I used in th' West to shoot th' fleet prairie dog, I climbed that precipitous ascent in th' face iv th' most gallin' fire I iver knew or heerd iv. But I had a few r-rounds iv gall mesilf an' what cared I? I dashed madly on cheerin' as I wint. Th' Spanish throops was dhrawn up in a long line in th' formation known among military men as a long line. I fired at th' man nearest to me an' I knew be th' expression iv his face that th' trusty bullet wint home. It passed through his frame, he fell, an' wan little home in far-off Catalonia was made happy by th' thought that their riprisintative had been kilt be th' future governor iv New York. Th' bullet sped on its mad flight an' passed through th' intire line fin'lly imbeddin' itself in th' abdomen iv th' Ar-rch-bishop iv Santiago eight miles away. This ended th' war."' . . .

"'I have thried, Hinnissy,' Mr. Dooley continued, 'to give you a fair idee iv th' contints iv this remarkable book, but what I've tol' ye is on'y what Hogan calls an outline iv th' principal pints. Ye'll have to r-read th' book ye'ersilf to get a thrue conciption. I haven't time f'r to tell ye th' wurruk Tiddy did in ar-rmin' an' equippin' himself, how he fed himsilf, how he steadied himsilf in battle an' encouraged himsilf with a few well-chosen wurruds whin th' sky was darkest. Ye'll have to take a squint into th' book ye'ersilf to l'arn thim things.'

"'I won't do it,' said Mr. Hennessy. 'I think Tiddy Rosenfelt is all r-right an' if he wants to blow his hor-rn lave him do it.'

"'Thrue f'r ye,' said Mr. Dooley, 'an' if his valliant deeds didn't get into this book 'twud be a long time befure they appeared in Shafter's histhry iv th' war. No man that bears a gredge again' himsilf 'll iver be governor iv a state. An' if Tiddy done it all he ought to say so an' relieve th' suspinse. But if I was him I'd call th' book "Alone in Cubia."'"

SOURCE: Finley Peter Dunne, *Mr. Dooley's Philosophy* (New York: R. H. Russell, 1900), pp. 13–18.

★ REMARKS BY "MR. DOOLEY" ★

- "A man that'l expict to thrain lobsters to fly in a year is called a loonytic; but a man that thinks men can be tur-rned into angels by an iliction is called a rayformer an' remains at large."
- "Thrust ivrybody—but cut th' ca-ards."

★ T.R. LEADS THE VOTERS TO SLAUGHTER ★

Teddy Roosevelt's Rough Riders were American heroes. So when T.R. ran for governor of New York in 1898 (immediately after the Spanish-American War), he had seven of his former soldiers campaign for him. New Yorkers turned out in droves to see the famed war heroes, but sometimes were disturbed by what the men had to say. At Port Jervis, New York, ex-sergeant Buck Taylor told the crowd:

"I want to talk to you about mah Colonel. He kept ev'y promise he made to us and he will to you. When he took us to Cuba he told us . . . we would have to lie out in the trenches with the rifle bullets climbing over us, and we done it. . . . He told us we might meet wounds and death and we done it, but he was thar in the midst of us, and when it came to the great day he led us up San Juan Hill like sheep to the slaughter and so he will lead you."

It is not known whether Buck was ever asked again to address a crowd, but Roosevelt was elected governor on November 5, and two years later became William McKinley's vice president.

SOURCE: William Henry Harbaugh, *Power and Responsibility: The Life and Times of Theodore Roosevelt* (New York: Farrar, Straus & Cudahy, 1961), p. 111.

★ MCKINLEY SEEKS POLITICAL HELP ★
FROM UP ABOVE

William McKinley may not be remembered as the Moses of his generation, but he was like Moses in one respect: he professed to receive instructions on how to govern his people directly from God. In 1899 he confided to a group of Methodist clergymen that his decision to annex the Philippines came after God had advised him to do it. "I walked the floor of the White House night after night until midnight," he said, "and I am not ashamed to tell you gentlemen, that I went down on my knees and prayed Almighty God for light and guidance more than one night." Finally, McKinley revealed, God heard his prayers and told him to take the islands as a gift from heaven.

SOURCE: Charles S. Olcott, *The Life of William McKinley* (Boston: Houghton Mifflin, 1916), II, 109–11.

★ SPEAKER REED SAYS TO BRING IN THE WASH ★

Speaker of the House of Representatives Thomas B. Reed was a brilliant parliamentarian, an incisive debater, and an extremely able party leader. But for none of these things was he famous. His reputation rested, rather, on his lightning-fast retorts and entertaining witticisms.

- "Once the House was making an effort to secure a quorum, and, as is usually done in such cases, telegrams were sent to members who were absent. One man, who was delayed by a flood on the railroad, telegraphed Reed, saying, 'Washout on line. Can't come.' Reed telegraphed back, 'Buy another shirt and come on next train.'"

- "[Reed] was bitterly opposed to our war with the Philippines, and he expressed his idea of the glory of the war in a concrete case in the following fashion. One morning, when the newspapers had printed a report that our army had captured Aguinaldo's young son, Reed came to his office and found his law partner at work at his desk. Reed affected surprise and said: 'What, are you working today? I should think you would be celebrating. I see by the papers that the American Army has captured the infant son of Aguinaldo and at last accounts was in hot pursuit of the mother.'"

- In the middle of one debate the note of impartiality was struck. Mr. Springer of Illinois was declaring with great solemnity that, in the words of Henry Clay, he would rather be right than be president. "The gentleman need not be disturbed," interrupted Reed, "he will never be either."

- When an irate Democrat came storming toward the podium of the Speaker of the House and demanded to know, "What becomes of the rights of the minority?" Reed casually glanced up and replied, "The right of the minority is to draw its salaries, and its function is to make a quorum."

- Speaker Reed: "A statesman is a successful politician who is dead."

SOURCES: Champ Clark, *My Quarter Century of American Politics* (New York: Harper & Brothers, 1920), I, pp. 290–91; Henry Cabot Lodge, *The Democracy of the Constitution* (1915; rpt. Freeport, N.Y.: Books for Libraries, 1969), p. 199; Edward Boykin, ed., *The Wit and Wisdom of Congress* (New York: Funk & Wagnalls, 1961), p. 165.

★ Four U.S. Reporters Start Boxer Rebellion ★

In 1899 a reporter from each of the major Denver, Colorado, newspapers, the *Republican*, the *Times*, the *Post*, and the *Rocky Mountain News*, met while covering a "slow" assignment in south Denver. Rather than return to their editors with simply another mundane and mediocre news item, the reporters decided to invent a truly newsworthy story.

Their first idea, to write that a Boston kidnapping victim was being held for ransom in Denver, was rejected because a quick phone call to the Boston police would expose the hoax too easily. Their story, they reasoned, must deal with foreign affairs. Only then would verification be difficult enough to ensure the story's acceptance. Finally the reporters agreed upon a story concerning the Great Wall of China. A party of engineers, they would report, had arrived in Denver from Wall Street en route to China. The engineers' company had negotiated with the Chinese government, which had decided to destroy the ancient boundary as a demonstration of its commitment to increased world trade. The engineers were traveling to the Far East to inspect the wall and to determine the costs of its demolition.

The next step was preparation. At the swank Windsor Hotel the four reporters signed fictitious names to the register and exacted a promise from the clerk to state, if anyone asked, that four "visitors" from the east coast had stayed at the hotel. These "visitors," the clerk was further instructed to say, had talked with the press, paid their bills, and gone west.

The reporters then returned to their respective newspapers and filed the "scoop." The sensational story ran page-one in all four of the Denver papers the next morning. By the end of the week east-coast newspapers had picked up the item. One of these papers even quoted a Chinese mandarin visiting New York City who confirmed the story. Being of such international interest, the story was cabled without comment to Europe and eventually to China itself.

In China the political situation was extremely unstable. Many Chinese, especially those affiliated with the "Boxer" athletic clubs, thought the government had bowed enough to the European imperialists. Now, they read, the government had contracted with a Western firm to destroy the Great Wall, monument to past Chinese glory. "It was the last straw," wrote missionary Bishop Henry W. Warren of the Methodist Episcopal Church, "and hell broke loose to the horror of the world."

Official denials of the story were ineffective. Xenophobia ran wild, missionaries were forced to flee, and foreign legations came under violent attack for weeks. The rebellion, which had been partly encouraged by the Chinese authorities to help curb the ambitions of foreigners, was now out of control. Even Emperor Kwang-su and his empress had to leave. Within a month, however, after a great loss of life and property, the Boxer Rebellion had been quashed and order restored.

SOURCE: Harry Lee Wilber, "A Fake That Rocked the World," *North American Review* (March, 1939), *passim*.

★ ADMIRAL DEWEY CATCHES POTOMAC FEVER ★

Admiral George Dewey had no experience in politics, but that did not matter. Because he was a military hero—he had won a great victory over the Spanish fleet in the Spanish-American War—he was automatically considered presidential material. Dewey said he was not interested in the office. "I am unfitted for it," he declared, "having neither the education nor the training." But his name kept being mentioned, particularly by Democrats who wanted him to wrest the nomination from William Jennings Bryan.

During the early months of 1900 reporters, eager for a big story, tried to get Dewey to declare for the presidency. But the admiral wouldn't.

Then one day in April, New York *World* reporter Horace J. Mock went by Dewey's house around six-thirty in the evening and asked the admiral the familiar question: Would he be a candidate? Without hesitating, Dewey answered, "Yes, I have decided to become a candidate."

The reasons? First, said Dewey, "If the American people want me for this high office, I shall be only too willing to serve them."

And second, "Since studying this subject, I am convinced that the office of the president is not such a very difficult one to fill."

Many Americans had hoped for months that Dewey would declare he was in the race. Now he finally had, but because of his comment "that the office of the president is not such a very difficult one to fill," no one wanted him anymore. Dewey had punctured his own rising balloon. The people didn't want a president who admitted he was up to the job only because the job was pretty low. Newspaper headlines read: "Leaders Laugh at Poor Dewey" and "The Entire Capital Is Laughing at the Former Hero."

SOURCE: Mark Sullivan, *Our Times* (New York: Scribner's, 1926), I, 309–38.

★ MCKINLEY ASSASSINATED INSTEAD OF A PRIEST ★

But for a few words spoken in an obscure Chicago tailor shop, William McKinley would never have been assassinated. In 1901 the man who would kill the President walked into the tailor shop of a friend and announced that after mature reflection on the state of society, he had decided to kill a priest. "Why kill a priest?" asked the friend. "There are so many priests; they are like flies—a hundred will come to his funeral."

The man reconsidered his decision and decided it would be better to assassinate a president. On September 6, 1901, Leon Czolgosz went to Buffalo, New York, and killed the President of the United States.

SOURCE: Mark Sullivan, *Our Times* (New York: Scribner's, 1930–36), II, 370–71.

★ MCKINLEY'S MURDER PROPHESIED ★

The assassination of President William McKinley at the Pan-American Exposition shocked the world, but did not come as a complete surprise to George B. Cortelyou. Cortelyou, the President's secretary, had worried that the trip to Buffalo might be dangerous. There would be, he felt, so many opportunities to attack the President. So, on his own authority, the secretary quietly removed from the agenda one of the trip's riskiest events: a reception for the public in which McKinley would shake hands with anyone who approached. Unfortunately, McKinley somehow learned about the cancellation and ordered that the reception be restored to the schedule.

Sometime before the event Cortelyou made one last attempt to persuade the President to cancel it. "Why should I?" McKinley asked him. "No one would want to hurt me." Cortelyou pointed out that the reception would not do the President much good. He would have only ten minutes to shake hands and could not possibly greet more than a small number of people. Thousands would be disappointed. But McKinley told Cortelyou: "Well, they'll know I tried, anyhow."

Cortelyou resigned himself to the decision. McKinley would have his shaking no matter what. The President loved handshaking and was quite expert at it. Reaching for a hand, he would grab it, give it one swift jerk, turn the visitor to the right, and then let go, ready for the next shake, all the while smiling broadly.

On September 5, President's Day, McKinley gave a speech and then engaged in fifteen minutes of unscheduled handshaking. Czolgosz was in the crowd, but that day didn't act.

Friday, September 6, began uneventfully. The President took a quiet trip up to Niagara Falls, had lunch at the International Hotel, and returned to Buffalo in midafternoon. At about three-thirty he proceeded to the Temple of Music for the handshaking reception that Cortelyou had tried to prevent.

At precisely four o'clock the doors of the temple were swung open and the public filed in. The handshaking started immediately. One by one the people stepped up to the President, extended their hand, and received the machine-gun McKinley grip. Forty-five people a minute shook the President's hand. After five or six minutes Cortelyou, anxious and cautious, ordered the door of the temple closed. A comely woman with a baby approached the President, followed by several other people. At about seven minutes past four Czolgosz, his right hand bandaged to conceal a gun, appeared before McKinley.

Two shots rang out, one deflected by a button, the other pene-trating the President's abdomen.

It had gone off exactly according to script—a script written by Czolgosz, a madman, and prophesied by Cortelyou, the President's secretary. On October 23, Leon Czolgosz was electrocuted for the murder of William McKinley.

SOURCE: Walter Lord, *The Good Years* (New York: Harper & Brothers, 1960), pp. 41–51.

★ HOW TO MAKE $36 MILLION WITHOUT ★ REALLY TRYING

Everyone knows that the great magnates of the Gilded Age fleeced the public. Yet in many cases the public literally jumped at the chance to be cheated by a Rockefeller or a Gould. Henry Rogers and William Rockefeller once earned $36 million without investing a dime, simply because the public was willing to buy anything the men's names were associated with, regardless of its value. The for-tune was made by "purchasing" the Anaconda Copper Company.

Rogers and Rockefeller gave a $39 million check to Marcus Daly for Anaconda, with the understanding that Daly would hold the check and not cash it in for a short time. Next, with their own clerks as dummy directors, the two robber barons founded the Amalgamated Copper Company. Amalgamated's first move? The new company, with no assets, printed up stock and randomly val-ued it at $75 million. With this paper the company "bought" Ana-conda from Rogers and Rockefeller. The two "entrepreneurs" then went to the bank and borrowed $39 million to cover their original check to Daly. The bank gladly agreed to use the Amalgamated stock as collateral. Next, of course, Rogers and Rockefeller sold the Amalgamated stock to the hungry public for $75 million cash. The bank loan was paid off with $39 million, leaving $36 million for the two partners. They also gained one copper company.

SOURCE: Robert L. Heilbroner, *The Worldly Philosophers* (New York: Simon & Schuster, 1953), pp. 202–3.

★ The Cure for Manure in the Streets ★

Every decade has a far-reaching cure for its ills. Around 1900 the problem was pollution. Three million horses inhabited urban America, with the healthier ones contributing from twenty to twenty-five pounds of manure each day. On every street their presence was evident as swarms of flies circulated and pungent odors permeated the air. To add to the atmosphere, almost every block boasted stables packed with urine-saturated hay.

Four-legged pollution was not alleviated by a change in the weather. When it rained, the streets turned to a muddy manure mush. During dry spells heavy carriage and foot traffic beat the dung to a fine dust which, as one contemporary put it, blew "from the pavement as a sharp piercing powder, to cover our clothes, ruin our furniture and blow up into our nostrils."

New York alone was home to approximately 150,000 horses or, pessimistically, to some ten million pounds of manure a year. The offerings of the 15,000 horses of Rochester, New York, in 1900, would have covered an acre of soil with a heap 175 feet high. In light of ever-increasing production, many Americans feared that their cities would soon disappear under the dung.

But a godsend from turn-of-the-century pollution was becoming available. At last, rejoiced Americans, the curtain was closing on the age of equine air. Cities would now be cleaner, quieter, healthier places in which to work and live. At last, the age of the automobile had arrived.

SOURCE: Otto L. Bettmann, *The Good Old Days: They Were Terrible* (New York: Random House, 1974), p. 3.

★ The Greatest Agent of Civilization ★

In his autobiography, *Up from Slavery*, Booker T. Washington wrote of his experiences as a teacher in a black school. Particularly, he emphasized the importance of nonacademic lessons: "I gave special attention to teaching [students] the proper use of the tooth-brush and the bath. In all my teaching I have watched carefully the influence of the tooth-brush, and I am convinced that there are few single agencies of civilization that are more far reaching."

SOURCE: Booker T. Washington, *Up from Slavery* (1901; rpt. Garden City, N.Y.: Doubleday, 1963), p. 54.

★ THE PRESS'S MISCOVERAGE OF T.R.'S ★ INAUGURATION

Following the death of William McKinley, several members of the press covered the brief swearing-in of the new president, Teddy Roosevelt. One New York reporter, particularly impressed with the ceremony, contrasted it in his article with a recent European coronation. The typesetter, however, mistakenly used the letter *b* instead of *o* in the word "oath." The next morning the paper's audience was informed: "For sheer democratic dignity, nothing could exceed the moment when, surrounded by the cabinet and a few distinguished citizens, Mr. Roosevelt took his simple bath, as President of the United States."

Not only were New York readers surprised to learn of their former governor's first act in office, but Europeans as well received a startling introduction to America's new president. The former Rough Rider took a presidential bath in the London papers, too.

SOURCE: Cleveland Amory and Frederic Bradlee, eds., *Vanity Fair Selections from America's Most Memorable Magazine* (New York: Viking, 1960), p. 293.

★ ADVICE ON THE PANAMA CANAL ★

Everyone knows that Teddy Roosevelt stole the Panama Canal. But even the man who swung the Big Stick thought he had to offer the world some sort of justification for his actions. Reasoning that seizure of the land would benefit "civilization," T.R. claimed that morally he had done the right thing. And because American actions were justified morally, the President argued, they were therefore justified legally. U.S. Attorney General Philander C. Knox offered the appropriate response to Roosevelt's legal opinion: "Oh, Mr. President, do not let so great an achievement suffer from any taint of legality."

After a painstaking rationalization of his actions in a 1903 cabinet meeting, T.R. demanded to know if his justifications would satisfy worldwide objections.

"Have I defended myself?" he asked.

"You certainly have," replied an all-knowing Elihu Root, secretary of war. "You have shown that you were accused of seduction and you have conclusively proved that you were guilty of rape."

SOURCE: Walter Lafeber, *The Panama Canal* (New York: Oxford University Press, 1978), p. 34.

★ T.R., A GREATER MAN THAN MOSES ★

"We were at Winston, North Carolina, during a term of court. The place where everybody forgathered in the evening was the old Parker and Jordan Tavern, now gone. The lobby or lounging room was a great big room about thirty feet by twenty, and there was a large fireplace at either end of the room.

"On this particular evening—it must have been in 1900 or along there—a number of us were sitting around talking about everything under the sun, when Mr. Pruden turned to me and said, 'Whom are you Republicans going to nominate for President this time, Mr. Meekins?' 'Theodore Roosevelt, I hope,' was my reply. Mr. Pruden was an admirer of Mr. Taft, whom he considered a great lawyer. Then he asked why I wanted Roosevelt. 'Because,' I said, 'I think Roosevelt is the biggest man in this country. I'd go further and say I think he is the biggest man in the world today.'

"Mr. Pruden said, 'Well, I expect you think he is the greatest man that ever lived.'

"I said there were few in history who were any greater, barring, of course, the Saviour.

"'Do you think he is a greater man than Moses was?'

"I said, 'Well, I don't know that I would say that. Moses was a wonderful leader.'

"And at that point another man in the crowd spoke up and

said, 'Well, I don't know if Teddy is a greater man than Moses, but I'll bet one thing—if Teddy had been leading the children of Israel I'll be durned if it would have taken 'em forty years to get out of the wilderness!'"

SOURCE: J. C. Meekins quoted by John G. Bragaw, "Random Shots," June 27, 1936, p. 11, in Manuscripts of the Federal Writers' Project of the Works Progress Administration for North Carolina, 1939, Archive of Folk Song, Library of Congress.

★ A GENTLEMAN GOES SWIMMING ★

Just because he was president and lived in metropolitan Washington, D.C., Theodore Roosevelt did not have to halt his wilderness explorations. Indeed, the chief of state would often take friends and government leaders on adventures through the marsh around the Potomac River. Jean Jules Jusserand, the ambassador from France, accompanied T.R. on many of these excursions.

Once, as the party was hiking on a particularly rocky trail, Jusserand used gloves to protect his hands. Farther downstream, as a relief from the hot Washington weather, the President suggested that the group take a swim. The idea seemed excellent, and the government leaders immediately stripped and jumped into the water. Jusserand, however, had not removed his gloves.

"Eh, Mr. Ambassador," asked the nude president, "have you not forgotten your gloves?"

Always the gentleman, the Frenchman looked down at his gloves. "We might meet ladies," he said.

SOURCE: Jean Jules Jusserand, *What Me Befell* (Boston: Houghton Mifflin, 1933), pp. 335–36.

★ T.R. TAKES A HUNDRED THOUSAND FROM ★ STANDARD OIL

A few weeks before the presidential election of 1904, Teddy Roosevelt learned that the Standard Oil Company had made a large contribution to his campaign chest. This worried the President, since he had repeatedly assailed Standard Oil as the worst of the one-eyed monster trusts. The Rough Rider freely accepted substantial contributions from other corporations—$150,000 from Morgan & Company, $100,000 from the Frick steel interests, $148,000 from New York Life—but this was different. In a letter to his campaign treasurer, released to the public on October 26, the President self-righteously announced: "I have just been informed that the Standard Oil people have contributed $100,000 to our campaign fund. This may be entirely untrue. But if true I must ask you to direct that the money be returned to them forthwith."

As T.R. was dictating this letter, his secretary of state happened to walk into the room. "Why, Mr. President," the secretary declared, "the money has been spent. They cannot pay it back— they haven't got it." To which Roosevelt replied, "Well, the letter will look well on the record, anyhow." And the letter was published.

In 1908, T.R. learned that the Standard Oil money had never been returned.

SOURCE: Matthew Josephson, *The President Makers* (New York: Harcourt, Brace, 1940), pp. 166–67.

★ THE BEST MAN FOR THE JOB ★

In 1906, Teddy Roosevelt named Oscar Straus, a Jew, as secretary of commerce and labor. At a dinner celebrating the appointment, T.R. explained that he had selected the new secretary without regard to race, color, creed, or party. The President had been accused by some of courting the large New York Jewish vote with the appointment. T.R. stressed, however, that his only concern had been to find the most qualified man in the United States for the job. Jacob Schiff, who was present at the dinner, was asked by the President to confirm this. Schiff, wealthy, respectable, old, and quite deaf, nodded emphatically and exclaimed, "Dot's right, Mr. President, you came to me and said, 'Chake, who is der best Jew I can appoint Secretary of Commerce?'"

SOURCE: John Morton Blum, *The Republican Roosevelt* (Cambridge, Mass.: Harvard University Press, 1961), p. 37.

★ "NUTHING ESCAPES MR. RUCEVELT" ★

On August 27, 1906, T.R. proved his willingness to risk controversy no matter how small the issue. Long a believer in the movement to simplify the spelling of words by eliminating silent vowels, he ordered the public printer to thenceforth use the simplified spellings of three hundred specified words in all government publications. Some of the spelling changes were minor, as in the words "honor," "parlor," and "rumor," and were subsequently adopted by dictionaries. But other changes did real violence to the old words and seemed to rob them of dignity and grace. "Kissed" became "kist," "blushed" became "blusht," "gypsy" became "gipsy." The *s* was dropped in favor of *z* in "artizan," "surprize," and "compromize," and the *e* dropped altogether in "whisky." Most disturbing of all at the time, curiously, was the change of "through" to "thru."

The simplified spelling movement had received the support of Nicholas Murray Butler, Andrew D. White, and David Starr Jordan, among educators, and had benefited from funding by Andrew Carnegie (he contributed more than $250,000). But the public was in no mood for spelling reform, and T.R. was assailed in all quarters for tampering with the language. Henry Watterson, editor of the Louisville *Courier-Journal*, wrote: "Nuthing escapes Mr. Rucevelt. No subject is tu hi fr him to takl, nor tu lo for him tu notis. He makes tretis without the consent of the Senit. He inforces such laws as meet his approval, and fales to se those that du not soot him. He now assales the English langgwidg, constitutes himself a sort of French Academy, and will reform the spelling in a way tu soot himself."

Congress was not in session when T.R. issued his edict, but when the members returned there was a controversial debate. After a short time Roosevelt was beaten, and a resolution was adopted ordering the Government Printing Office to "observe and adhere to the standard of orthography prescribed in generally accepted dictionaries of the English language."

SOURCE: Mark Sullivan, *Our Times* (New York: Scribner's, 1930–36), IV, 162–90.

★ T.R. USES SECRET SERVICE TO BLACKMAIL ★ CONGRESSMEN

True or not, the story was believed to be true by a great many congressmen. And that was what mattered.

The story was that President Roosevelt had recently become a blackmailer. Concerned about several important votes that would be close, he had dispatched the Secret Service to dig up dirty information about members of Congress. Specifically, he had ordered the service to find out which members frequented the capital's whorehouses. Then he had used this information to force the members in question to adopt the administration's line on key votes—or else!

There was no proof to the story, but it did not seem entirely farfetched either. During the previous session Congress had passed an amendment requiring the Secret Service to limit its investigations to the activities of the executive branch. T.R. was now asking that the amendment be repealed. The President openly stated that the Secret Service ought to be permitted to investigate anyone—even members of Congress. Did T.R. want a private police force to blackmail congressmen? Some members thought he did.

Clearly, most congressmen did not believe the charges against Roosevelt. But the story created tension between the President and Congress. Indirectly it made congressmen extremely sensitive about the independence of their branch of government. What did Roosevelt mean when he said the service should be allowed to investigate congressmen? Was he implying that there were congressmen who had committed crimes? Was every congressman under suspicion of having broken the law?

Eventually the legislators became so worked up about Roosevelt's insinuations that they decided to censure him—just two months before the end of his term. It was only the second time in history (excluding the impeachment of Andrew Johnson) that the Congress censured a president. The first time had been when the

Congress censured Andrew Jackson during the controversy over the Bank of the United States. Jackson's censure, however, was eventually expunged from the records of Congress through the efforts of Thomas Hart Benton. The censure of Theodore Roosevelt was never removed.

Source: Henry F. Pringle, *Theodore Roosevelt* (New York: Harcourt, Brace, 1931), pp. 483–85.

★ Truly American Names ★

In 1903 an order was issued by the Commissioner of Indian Affairs requiring all Indians on reservations to adopt plain American names. The commission strongly recommended that the Indians use patriotic names. Americans at the time speculated upon the changes this would make in the news items issued from the reservations: "No longer will 'Tail Feathers Coming' woo dusky 'Minnie Weeping Willow'; no longer will 'Blanket on Straight' journey to the portals of the sunset and on the way stop at the wigwam of 'Two Bones,' the ancient arrow-maker."

Americans suggested that news items of the future might read: "Patrick Henry has sold his interest in the Custer shooting gallery to Abraham Lincoln. Daniel Boone and Martin Van Buren have been sentenced to two days in jail by Police Judge John Paul Jones. George Washington and James Fenimore Cooper will fight tonight at the Sports Club."

Source: Homer Croy, *What Grandpa Laughed At* (New York: Duell, Sloan & Pearce, 1948), p. 103.

★ Huck Finn Banned ★

When Mark Twain was told in 1905 by the librarian of the Brooklyn Public Library that copies of *The Adventures of Tom Sawyer* and *The Adventures of Huckleberry Finn* had been removed from the shelves of the children's room, he replied, "I wrote *Tom Sawyer* and *Huck Finn* for adults exclusively, and it always distresses me when I find that boys and girls have been allowed access to them. The mind that becomes soiled in youth can never again be washed clean."

SOURCE: Anne Lyon Haight, *Banned Books*, 3d ed. (New York: R. R. Bowker, 1970), p. 57.

★ THAT FIRST AIRPLANE ★

On Thursday, December 17, 1903, Wilbur and Orville Wright made history by making a plane fly for fifty-nine seconds. But no one seemed to notice.

Printer's ink flowed at the Norfolk *Virginia-Pilot*, but virtually nowhere else. Not in the Wright brothers' hometown newspaper, the Dayton *Journal*, and not in the big city papers. The story did appear in the New York *Tribune*—in the sports section.

The muffled reaction had many causes. Partly it was due to disbelief. A noted professor, after all, had just published an article packed with charts and diagrams proving that man could never fly. Sure, the Wright brothers had shown that a plane could fly—but only for about a minute, which was hardly page-one news. Santos-Dumont had demonstrated that a dirigible could stay up in the air for nearly an hour. Then there was the belief that even if a plane could fly for longer periods of time, it would not be able to carry cargo. And virtually no one thought that planes would ever carry passengers. Finally, the silent reaction was due to the Wright brothers themselves. At bottom they were mechanics and inventors, not showmen. They did not have the flair of a Charles Lindbergh, or the shrewdness of a P. T. Barnum. So they had trouble persuading people of the possibilities of their contraption.

The Wright brothers believed that one of the chief uses for the airplane would be to maintain international peace. With airplanes nations could keep an eye on one another and avoid sudden wars. The brothers would undoubtedly have been horrified to learn that their invention made war worse than it had ever been.

Portents of the future came in 1905, when British lieutenant colonel J. E. Clapper expressed interest in the Wright brothers' airplane. Whether for peaceful or military purposes he did not say. But negotiations fell through. In the meantime the Wright brothers furiously tried to attract the attention of Washingon. It would be far better for their own government to make use of the airplane than a foreign one. But Washington ignored them. Finally, however, just when the British offer was withdrawn, Washington came to its senses. The year was 1907. It had taken four years for the government to realize the importance of the airplane.

SOURCE: Walter Lord, *The Good Years* (New York: Harper & Brothers, 1960), pp. 97–100.

★ HENRY JAMES ASKS DIRECTIONS ★

Henry James loved motoring, but did not have a sense of direction. Neither did Edith Wharton. Here Wharton describes what happened when she and James, on a trip in England, arrived in Windsor late one night:

"We must have been driven by a strange chauffeur—perhaps Cook was on a holiday; at any rate, having fallen into the lazy habit of trusting to him to know the way, I found myself at a loss to direct his substitute to the King's Road. While I was hesitating, and peering out into the darkness, James spied an ancient doddering man who had stopped in the rain to gaze at us. 'Wait a moment, my dear—I'll ask him where we are'; and leaning out he signalled to the spectator.

"'My good man, if you'll be good enough to come here, please; a little nearer—so,' and as the old man came up: 'My friend, to put it to you in two words, this lady and I have just arrived here from *Slough*; that is to say, to be more strictly accurate, we have recently *passed through* Slough on our way here, having actually motored to Windsor from Rye, which was our point of departure; and the darkness having overtaken us, we should be much obliged if you would tell us where we are now in relation, say, to the High Street, which, as you of course know, leads to the Castle, after leaving on the left hand the turn down to the railway station.'

"I was not surprised to have this extraordinary appeal met by silence, and a dazed expression on the old wrinkled face at the window; nor to have James go on: 'In short' (his invariable prelude to a fresh series of explanatory ramifications), 'in short, my good man, what I want to put to you in a word is this: supposing we have already (as I have reason to think we have) driven past the turn down to the railway station (which, in that case, by the way, would probably not have been on our left hand but on our right), where are we now in relation to . . .'

"'Oh, please,' I interrupted, feeling myself utterly unable to sit through another parenthesis, 'do ask him where the King's Road is.'

"'Ah—? The King's Road? Just so! Quite right! Can you, as a matter of fact, my good man, tell us where, in relation to our present position, the King's Road exactly *is?*'

"'Ye're in it,' said the aged face at the window."

Source: Edith Wharton, *A Backward Glance* (New York: D. Appleton-Century, 1934), pp. 242–43. Reprinted by permission of Herbert A. Fierst on behalf of William R. Tyler.

★ Death by Urination ★

Pat Garrett was the lawman who killed Billy the Kid in 1881. For that one feat, Garrett's fame spread throughout the Territory of New Mexico and the nation. In 1908, Garrett died one of the Old West's most embarrassing deaths. With Carl Adamson, the old lawman was riding in a buggy down a lonely southeastern New Mexico road, while Wayne Brazel, another acquaintance, followed behind. Suddenly, Adamson pulled over to the side of the road and jumped out to urinate. Garrett followed suit, carefully holding his shotgun in his right hand to ensure that it would not accidentally fire. He removed his left glove, unbuttoned his trousers, and began urinating. Before he had finished, a bullet slammed into the back of his head. Garrett died immediately.

Incredibly, no one was ever convicted of the murder of Pat Garrett. Wayne Brazel, who confessed to the killing as "self-defense," was halfheartedly tried and found not guilty. Carl Adamson was not even called to court as a witness. Apparently, the reason for all this was that Pat Garrett had few friends in the area after killing Billy the Kid, a popular folk hero in New Mexico. Differing New Mexico legends connect at least five people, Brazel and Adamson included, with the murder.

SOURCE: Leon C. Metz, *Pat Garrett* (Norman, Okla.: University of Oklahoma Press, 1974), p. 291.

★ ONE WAY TO KILL A FILIBUSTER ★

Robert M. LaFollette was one of those politicians who made the blood of robber barons boil. He wouldn't deal, would not sacrifice his independence or his principles. He was that worst kind of politician, the kind that could not be bought.

On May 29–30, 1908, the Wisconsin senator was leading a filibuster against the Aldrich-Vreeland bill. Aldrich-Vreeland was a perfectly legitimate bill, designed to allow the currency to expand during times of panic. But for some reason LaFollette disliked it.

He began his filibuster at twenty minutes past twelve on the afternoon of May 29 and kept it up hour after hour. Four o'clock, six o'clock, nine o'clock. Every so often, to maintain his energy, he would drink a mixture of milk and raw eggs prepared for him by the Senate restaurant. Sometime between ten and eleven o'clock he received one of these mixtures and began drinking it. Suddenly his face twitched and took on a sour expression. He stopped drinking immediately. The thought occurred to him that someone was trying to poison him. Perhaps one of his many enemies had finally decided it was time to be rid of Robert M. LaFollette. Surely he had given them cause. He ordered the mixture taken away and analyzed.

A little later LaFollette, still leading the filibuster, began to feel queasy. Soon he came down with the painful symptoms of dysentery, but still he would not leave the Senate. Heroically he stayed on, for another half dozen hours. Finally, he gave up, at precisely 7:03 in the morning.

LaFollette had kept talking for sixteen hours and forty-three minutes—the longest filibuster in history up to that time. But it was not long enough. Later that day Aldrich-Vreeland was passed.

Sometime afterwards a laboratory reported on the mixture that had made LaFollette sick. According to chemical analysis the mixture was indeed full of poison—ptomaine—enough to kill a man. Big Money or someone had tried to kill a filibuster by murdering the filibusterer. The figurative use of the phrase "kill a filibuster" had taken on a haunting, literal meaning. No one was ever arrested for the attempted murder.

SOURCE: U.S. Senate Historical Office, notes.

★ A ROCKEFELLER TAKES THE FIFTH ★

Heard increasingly during investigations of big business in the late nineteenth century was the refrain "I refuse to answer on the advice of counsel." In a railroad rate case involving William Rockefeller, brother of the founder of Standard Oil, repeated use of the phrase provoked a ludicrous exchange:

"On the ground that the answer will incriminate you?"

"I decline to answer on advice of counsel."

"Or is it that the answer will subject you to some forfeiture?"

"I decline to answer on the advice of counsel."

"Do you decline on the ground that the answer will disgrace you?"

"I decline to answer on the advice of counsel."

"Did your counsel tell you to stick to that one answer?"

"I decline to answer on the advice of counsel."

With that the whole room burst into laughter, Rockefeller included.

SOURCE: Frederick Lewis Allen, *The Big Change* (New York: Harper & Brothers, 1952), pp. 71–72.

★ WHEN SOLAR ENERGY FLOURISHED ★

Solar energy is not new to America. Home solar heating began in this country at the start of the twentieth century, when it was introduced in Florida and California. Most of the units were simple devices built by amateurs, and consisted of a single sheet of glass, a metal box, and some copper tubing. But there were a few companies in the field. The Day/Night Company, for example, began selling solar heating units in the first years of the century.

By the 1930s there were tens of thousands of units operating in the United States, but the prospects for solar energy were dim. As a professor lamented in 1936, few textbooks on home heating even mentioned solar energy. The professor proved solar heating was economical, but most people preferred the ease of the increasingly popular gas heater. A few solar energy firms, including the pioneering Day/Night Company, switched in the 1930s to the manufacture of gas heaters.

By the 1950s there were still many solar-heated homes—over 50,000 in Miami alone. But by the 1970s there were hardly any—a fact which caused many people to believe that solar energy was the untried alternative to conventional energy.

Source: Wilson Clark, *Energy for Survival* (Garden City, N.Y.: Anchor, 1975), pp. 370–74.

★ Hollywood's Ascendancy ★

Contrary to popular belief, Hollywood's good weather had little to do with its establishment as the movie capital of the United States. It was chosen not because of its warm climate but primarily because it lay within easy reach of the Mexican border.

In the 1910s the movie industry was based in New York and was completely dominated by a single trust. The Motion Picture Patents Company controlled through its patents virtually every phase of the business, from production to distribution. No one could legally make a movie without its permission. Anyone who tried faced severe penalties.

Still, there were people who wanted to make films on their own. These people did not want to be hampered by the trust's cautious approach to filmmaking. So they left New York and moved to areas where they could not easily be prosecuted by the trust. Some went to Cuba to escape the law completely, while others went to Florida and Los Angeles. But Cuba was threatened by disease and Florida proved too hot. Within a short time everyone had moved to Hollywood. The new movie capital was everything a producer could want; it had good weather and cheap labor. Most important, it was close enough to the Mexican border to afford quick escapes from the law in times of legal trouble.

SOURCE: Lewis Jacobs, *The Rise of the American Film* (New York: Harcourt, Brace, 1939), pp. 81–86; lecture by Donald Fleming at Harvard University, Spring 1978.

★ A SUPREME COURT JUSTICE'S LAST WORDS ★

"Shortly before his death [Supreme Court Justice John Marshall Harlan] became partly conscious and spoke his farewell words to those who were at his bedside, evidently with great difficulty: 'Good-by, I am sorry to have kept you all waiting so long.'"

SOURCE: New York *Sun*, October 15, 1911.

★ SIXTY YEARS BEFORE WATERGATE ★

In 1912 no one had ever heard of wiretapping; the word wasn't even invented until after World War I. But the practice itself was known. At the Republican convention in Chicago someone tapped the long-distance phones used to keep Teddy Roosevelt, who was home in Oyster Bay, New York, in touch with his managers. It was the first known case of wiretapping in American politics. When T.R. learned about it, he went to Chicago so that he could direct his managers in secret.

SOURCE: William Roscoe Thayer, *Theodore Roosevelt* (New York: Grosset & Dunlap, 1919), p. 359.

★ T.R.—SAVED BY THE SPEECH ★

It is not always bad that politicians write long speeches. In 1912, when Teddy Roosevelt was campaigning for the presidency as the candidate of the Bull Moose party, he prepared his speeches on small sheets of paper with extra spacing between the lines for easier delivery. When the Rough Rider was in Milwaukee on October 14, his speech covered fifty heavy, glazed pages. Folded over and carried in his pocket, they numbered one hundred. This was a fortunate length, for as Roosevelt left his hotel for the rally, a would-be assassin fired a shot directly at the Bull Moose's heart. The bullet traveled through Roosevelt's coat, vest, eyeglass case, and the hundred-page speech and lodged against his fifth rib, cracking it, but not badly injuring the ex-president. Had the speech not altered the course and speed of the bullet, the missile would have passed directly through Roosevelt's heart and killed him.

The old Rough Rider, after realizing that his wound was only superficial, leaped at the chance to turn a near disaster to his political profit. For the past few days Roosevelt had been ill. Several appearances had even been canceled. That night in Milwaukee he had planned to say only a few words to the crowd, while an assistant read the bulk of his speech for him. But now, with a bullet lodged harmlessly in his chest and bloodstains showing on his shirt, he changed his plans.

"This is my big chance," he exclaimed to his physician, "and I am going to make that speech if I die doing it."

Once onstage, Roosevelt immediately informed the audience of his traumatic experience, boasting, "But it takes more than that to kill a bull moose!"

A deep sigh erupted as the ex-president pulled the speech from his coat pocket, with bullet holes in plain view of all. Later, another gasp filled the air as T.R. unbuttoned his vest and revealed his bloodied shirt. Although suffering, Roosevelt personally delivered all fifty pages of his speech.

Unfortunately, T.R.'s heroics were not enough. Not only did he fail to carry the national vote in 1912, he did not even carry Wisconsin.

SOURCE: Oscar King Davis, *Released for Publication* (Boston: Houghton Mifflin, 1925), 375–86.

★ T.R. LAMENTS PASSING OF THE GREAT RACE ★

Teddy Roosevelt was a great many things, but he was not a liberal on race questions. He did risk controversy once by inviting Booker T. Washington to eat dinner with him at the White House, but the President did not believe in racial equality.

That was made clear by his endorsement of one of the most appallingly racist books ever published, Madison Grant's *Passing of the Great Race*. Grant's theme was a familiar one: that unrestricted immigration into the United States had weakened the fabric of society and undermined the security of whites. But his solutions to the problems immigrants had supposedly caused were novel and extreme. In addition to recommending stiffer laws against miscegenation, he asserted that the inferior races ought to be sterilized and unfit individuals put to death. As he explained at the beginning of the book: "Mistaken regard for what are believed to be divine laws and a sentimental belief in the sanctity of human life tend to prevent both the elimination of defective infants and the sterilization of such adults as are themselves of no value to the community. The laws of nature require the obliteration of the unfit, and human life is valuable only when it is of use to the community of race."

Here is how Teddy Roosevelt described Grant's book in a blurb on the volume's jacket:

Revised, with Documentary Supplement

THE PASSING OF THE GREAT RACE

MADISON GRANT

"The book is a capital book—in purpose, in vision, in grasp of the facts our people must need to realize. It shows an extraordinary range of reading and a wide scholarship. It shows a habit of singular serious thought on the subjects of most commanding importance. It shows a fine fearlessness in assailing the popular and mischievous sentimentalities and attractive and corroding falsehoods which few men dare assail. It is the work of an American scholar and gentleman; and all Americans should be sincerely grateful to you for writing it."

—*Theodore Roosevelt.*

SOURCE: Madison Grant, *The Passing of the Great Race*, 4th ed., rev. (New York: Scribner's, 1930), p. 49.

★ THE ORIGIN OF FROZEN FOODS ★

Clarence Birdseye, the founder of the corporation that bears his name, was the discoverer of the frozen-food process. In 1912 the Massachusetts naturalist, on a fur-trading expedition to Labrador, happened to be fishing through the ice on a day when the temperature had dropped to a blistering twenty degrees below zero. The temperature was so cold that the fish he caught froze solid instantly when removed from the sea. Birdseye took his frozen fish back to camp and casually tossed one of them in a bucket of plain water. Miraculously, before Birdseye's eyes, the fish revived and began to dart left and right. The naturalist did not know quite what to make of this remarkable resurrection, but he watched intently. After several years of hard thought, he finally concluded that the fish had survived because it had been frozen quickly. Birdseye then reasoned that food could be preserved the same way. And in 1925 he marketed the first frozen food—fish.

SOURCE: Daniel Boorstin, *The Americans: The Democratic Experience* (New York: Random House, 1973), pp. 330–31.

★ BEFORE THE *TITANIC* ★

The sinking of the *Titanic* on a cold April night in 1912 in the icy waters of the Atlantic stunned the world. Partly the shock was due to the enormous loss of life: 1,500 dead. Partly it was because the British luxury liner was loaded with rich and glamorous passengers. So many, as a matter of fact, that not a few scheduled high-society events were canceled or postponed that spring. But mostly, perhaps, the world was startled because it had had so much faith in the new ship. The largest ship in the world . . . declared unsinkable by experts . . . so safe it carried just a few lifeboats. A ship that symbolized the good and expensive life.

But the sinking of such a ship had not been unanticipated.

Fourteen years before, a young American named Morgan Robertson had written a novel about a similar ship, filled with fabulously wealthy passengers, which hit an iceberg in the Atlantic one cold April night and went down—a ship named *Titan*.

Robertson's ship was remarkably like the real one. Both vessels were triple-screw and could reach about twenty-five knots. The real ship was 800 feet long; Robertson's 882.5 feet. The *Titanic* weighed about 66,000 tons fully loaded; the fictional ship weighed about 70,000 tons fully loaded. Both had a maximum capacity of approximately 3,000 passengers, and neither was equipped with an adequate number of lifeboats. Both ships were reputedly unsinkable; both were the largest ships in the world.

In the novel the big ship represented the best of modern society, just as the *Titanic* would fourteen years later. And, of course, the sinking of the *Titan* shocked the civilized world as much as the sinking of the *Titanic*.

The name of Robertson's novel was *Futility*.

SOURCE: Walter Lord, *A Night to Remember* (New York: Holt, Rinehart & Winston, 1955), p. 9.

★ WOODROW WILSON, POLICEMAN ★

Woodrow Wilson always loved automobile rides. But after his stroke he became greatly disturbed about drivers who speeded—or drivers who seemed to be speeding. The President had ordered that his car never be driven faster than fifteen or twenty miles an hour. But he believed that anyone who passed him had to be going at a dizzying, reckless rate. Finally he decided to do something about such people. He ordered the Secret Service to begin pursuing cars that passed the presidential limousine so that the drivers could be hauled back for questioning. The agents would take off after the offending drivers, but they always returned empty-handed, saying they had been unable to overtake the speeder.

Wilson, in the meantime, considered what could be done to the speeders if they were caught. He wrote a letter to the attorney general asking whether presidents had the powers of a justice of the peace. Wilson told the Secret Service that if the attorney general answered him affirmatively, the President would begin arresting speeders himself and trying them right on the road. There would be justice on the streets of Washington.

But not one speeder was ever apprehended. Eventually the Secret Service persuaded Wilson that it would be inappropriate for a president of the United States personally to try the cases of speeding drivers.

SOURCE: Gene Smith, *When the Cheering Stopped* (New York: William Morrow, 1964), pp. 146–47.

★ HUMOR BY WOODROW WILSON ★

Woodrow Wilson had a subtle sense of humor. On July 1, 1912, Wilson passed Champ Clark in the voting for the presidential nomination for the Democratic party. It was the thirtieth ballot. A reporter interrupted Wilson in a conversation with his family to ask the New Jersey governor for a statement. Wilson, emotionless concerning the good news, continued telling his family a favorite limerick:

There was a young lady from Niger
Who smiled as she rode on a tiger
They came back from the ride
With the lady inside
And the smile on the face of the tiger.

The reporter, not satisfied with the front runner's comment, begged the future president for a more direct, "excited" statement.

"You might put it in the paper," Wilson said with a quiet

smile, "that Governor Wilson received the news that Champ Clark had dropped to second place in a riot of silence."

SOURCE: Arthur Link, ed., *The Papers of Woodrow Wilson* (Princeton: Princeton University Press, 1977), XXIV, 516.

★ A KLUNKER FOR A KLANSMAN ★

On June 26, 1916, Joseph Simmons, Imperial Wizard of the Ku Klux Klan, issued the following "Imperial Decree" from his "Aulic in the Imperial Palace in the Imperial City of Atlanta": "The Kloran is *the* book of the Invisible Empire and is therefore a sacred book with our citizens, and its contents must be rigidly safeguarded. The book or any part of it *must* not be kept or carried where any person of the 'alien' world may chance to become acquainted with its sacred contents as such. *In warning:* A penalty sufficient will speedily be enforced for disregarding the decree in the profanation of the Kloran."

Six months later Simmons decided that a book as important as the Kloran should be officially recognized, so he applied to Washington for a copyright. Like any author, he forwarded one dollar and two copies of the book to the Register of Copyrights. And from that time forth *The Book of the Invisible Empire* was available to anyone who asked for it at the Library of Congress.

SOURCE: Laurence Greene, *The Era of Wonderful Nonsense* (Indianapolis: Bobbs-Merrill, 1939), pp. 86–87.

★ V. I. LENIN AND AN AMERICAN ROMEO ★

Late in the afternoon of April 11, 1917, V. I. Lenin made a telephone call to the American legation at Berne, Switzerland. Just a few days earlier the Bolsheviks had overthrown the czar. Now Lenin, a key leader of the revolutionaries, was trying to deliver an important message to the Americans concerning World War I. Lenin liked Woodrow Wilson and wanted to help the United States.

Over the telephone Lenin gave his name and explained that he would be arriving in Berne late in the day and would have to speak with someone at the legation. He had extremely significant news. Lenin was asked if he could not wait until the next day and come during official hours. No, he answered emphatically, "tomorrow will be too late. I must talk to someone this afternoon. It is most important. I must see someone."

But Lenin would see no one. His thick German accent marked him as just another émigré, maybe even a crank. His purported news? Well, it could surely wait until the following day. In any case, the man at the other end of the line, the duty officer at the legation, did not want to wait. He had a date with a pretty woman, a woman who had scorned him until recently as too young. He certainly could not break his evening engagement with her. So Lenin was told: "I'm sorry, it will have to be tomorrow. Ten o'clock tomorrow, when the office opens."

Six days later V. I. Lenin, the insignificant émigré, announced to the world the message he had wanted to tell the United States: Russia was taking itself out of the Great War. Peace negotiations with Germany were to begin promptly.

The name of the duty officer who refused to meet with Lenin because of his date was Allen Dulles, then a green intelligence man, later head of the Central Intelligence Agency.

SOURCE: Leonard Mosley, *Dulles* (New York: Dial Press/James Wade, 1978), pp. 46–48.

★ MILLARD FILLMORE AND THE BATHTUB ★

On December 28, 1917, H. L. Mencken published in the New York *Evening Mail* a startling account of the history of the bathtub in America. The history was a complete fabrication from beginning to end, and contained utterly fantastic claims, but people across the country believed it entirely. The story began to show up in other newspapers, and eventually creeped into works of scholarship. To this day it can be found in respected histories of the United States.

Mencken's story began with the patently improbable claim that the bathtub was unknown in America until 1842, when Adam Thompson, a Cincinnati cotton and grain dealer, recently back from a trip to Europe, where he had laid eyes on a bathtub for the first time, introduced it amid much fanfare. Unfortunately, the celebration did not last long. Thompson, the hick from Cincinnati, could not persuade people in the cosmopolitan East of the bathtub's usefulness. Instead, the city slickers decided that the bathtub was Public Enemy Number One. Eminent physicians expressed the opinion that bathing posed a serious danger to health, while legislators in the nation's two most advanced cities—Philadelphia and Boston—passed laws against it. The Boston town fathers decreed that no one could take a bath without the advice and consent of a doctor.

According to the story, opposition to the bathtub remained strong until 1851, when Millard Fillmore—described as "intrepid"—ordered one installed in the White House. After that bathing became fashionable.

For ten years Mencken watched in silent disbelief as his ridiculous story passed the lips of millions of gullible Americans. Finally, in May 1926, he revealed that the whole purported history was a pack of lies that he had fabricated in 1917 "to sublimate and so make bearable the intolerable libido of the war for democracy." He had not expected, he wrote, that anyone outside of a few raving idiots would believe it.

About thirty newspapers, reaching almost 250 million people, printed Mencken's confession of the bathtub hoax. But to his astonishment the story would not die down. That June the Boston *Herald,* which had published Mencken's disclosure three weeks before under the title "The American People Will Swallow Anything," published the original story as a news item.

Again Mencken wrote that the whole thing was a fake, but to no avail. A few months later *Scribner's Magazine* printed the old story as fact. In the 1930s someone wrote a whole book based on Mencken's spoof, and in the early 1970s a prize-winning historian related the discredited facts in his widely acclaimed trilogy on the American experience. In the middle 1970s the story made its way into the comprehensive multivolume *Dictionary of American History*.

SOURCE: James T. Farrell, ed., *H. L. Mencken: Prejudices* (New York: Vintage, 1955), pp. 242–47.

★ WHEN AMERICANS CELEBRATED THE ★ END OF WORLD WAR I

On the afternoon of November 7, 1918, Americans across the country learned that the "war to end all wars" had ended. Instantly offices and factories closed and people poured into the streets to celebrate. A sign at Rogers Peet Company read: "Who can work on a day like this? Gone to celebrate. Open tomorrow." Dozens of other stores blazoned the same message across their tightly shut doors. New York City hosted one of the largest ticker-tape parades in its history. For the first time since 1914 the world was at peace.

Or so the newspapers reported. Actually, heavy fighting was continuing along all fronts. There had been no armistice. Only in the United States and in Brest, France, did people believe that the war had come to a close.

On the afternoon of November 7, Admiral Henry B. Wilson, director of naval operations in French waters, received at his headquarters in Brest a message from Paris that the war had ended. The message, perhaps the work of German spies, was completely false. But Wilson believed it. He had little reason not to. Everyone knew that the war was almost over.

At a four-o'clock appointment with Roy Howard, president of United Press, Wilson casually revealed the message and told Howard he was free to use it. Howard, of course, leaped at the opportunity. He would be the first person to send the news of peace to the United States. The story would make his reputation.

It was only by chance that Howard was in Brest that afternoon. He had arrived in the city early that morning from Paris to catch a ship leaving immediately for the United States. But a friend had told him about another ship which would reach the United States sooner, though it would not leave until the next day. Howard had decided to take the later ship. That afternoon, having nothing to do, he had gone to see Wilson, who happened to be the kind of person who liked talking with reporters.

After Howard received the news of peace, he immediately had the message typed up at Brest's daily newspaper, *La Dépêche*. Then he raced over to the French cable office to send the news on to the United States.

Ordinarily, nothing would be sent out of the cable office without first being approved by French censors. But Howard's message had been typed on the *La Dépêche* machine that was used to receive wires from Paris—wires approved by the Paris censors and thus never questioned by the Brest cable operators. In addition, the cable was signed "Howard-Simms." Simms was the UP reporter whose name usually appeared on the Paris dispatches. Howard had cosigned the cable with the reporter's name to give Simms some of the credit for the greatest scoop of the century. So Howard's message was reported out of Brest completely uncensored.

When the cable arrived in the United States, there was no reason to disbelieve it. The story had to be true for it to have got past the French censors. It was immediately sent out across the entire country.

That night Howard learned, while sitting in a bar in Brest, that his scoop was completely false. At first Wilson encouraged him to believe that the story was simply "premature." But as the night lengthened, it became clear that the story was plainly incorrect. The scoop of the century had become the hoax of the century.

The next morning news of Howard's mistake was splashed across the front pages of every paper in America. Wilson took responsibility for the mistake, but everyone blamed Howard— blamed him for disappointing millions of people and for needlessly causing the nation's offices and factories to close down. One New York paper argued that Howard should be forced to pay for cleaning up the ticker tape that cluttered every street.

Three days later it was officially declared that the war was over. Howard had been wrong by only a few days.

SOURCE: Arthur Hornblow Jr., "The Amazing Armistice," *Century Magazine*, CIII (November 1921–April 1922), pp. 90–99.

★ WOODROW WILSON'S BARBER ★

After his stroke President Wilson grew a long beard, which his doctors sometimes considered shaving. On one occasion, standing by Wilson's bed, they were discussing the matter when one of them suggested that a doctor could shave off the beard. "You know," the man said, "in the olden days the doctors were barbers. Doctors were really barbers in those days." There was a cry from the bed: "They are barbarous yet."

SOURCE: Gene Smith, *When the Cheering Stopped* (New York: William Morrow, 1964), p. 109.

★ VICE PRESIDENTS ARE HELPLESS ★

Thomas R. Marshall, vice president under Woodrow Wilson: "The Vice President of the United States is like a man in a cataleptic state: he cannot speak; he cannot move; he suffers no pain; and yet he is perfectly conscious of everything that is going on about him."

Marshall also believed that the vice president was like an animal in a cage. When visitors to the Capitol peered at him in his office, Marshall would sometimes blurt out, "If you don't come in, throw me a peanut."

SOURCES: Holman Hamilton, *White House Images and Realities* (Gainesville, Fla.: University of Florida Press, 1958), pp. 4–5; Henry Graff, "A Heartbeat Away," *American Heritage*, August 1964, p. 86.

★ RELEASED FROM HIS CATALEPTIC STATE ★

On November 23, 1919, Vice President Marshall was in Atlanta speaking in the civic auditorium when a policeman suddenly rushed up to the podium and began talking in hushed tones to a local official. Marshall was then interrupted and given the terrifying news: according to a phone call just received, the President had died. Marshall staggered a few feet, then took hold of himself firmly. He announced to the audience: "I cannot continue my speech. I must leave at once to take up my duties as Chief Executive of this great nation." He asked for the people's prayers. Then, surrounded by a police escort, he left for his hotel.

"A most cruel hoax," was the way the Vice President described the incident when he discovered that Wilson was very much alive. Marshall had been "president" for barely an hour. Instantly the embarrassed Vice President slipped back into the comfortable anonymity to which he had grown accustomed. A short time later he left Atlanta by train, unescorted, and cheered only by the shrieking wheels of the locomotive.

SOURCE: Gene Smith, *When the Cheering Stopped* (New York: William Morrow, 1964), pp. 127–28.

★ WILSON'S BID FOR A THIRD TERM ★

At the end of his second term Woodrow Wilson was a broken man. The victim of a stroke that paralyzed much of his left side, he could barely move. He had, it is true, improved. Immediately after his stroke he was unable to move at all and could barely utter a single word. Now, at least, he had spasms of articulateness, lasting anywhere from a week to ten days. But Wilson was still very feeble. He could hold only brief conversations, often fell into deathlike sleep at the movies, and showed signs of creeping ludicrousness. He was so weak that he never held cabinet meetings, never entertained, and only once mustered enough strength to see a Republican senator.

But Woodrow Wilson wanted a third term. A third term would vindicate the Versailles Treaty, the League of Nations, and Wilson himself. Most of his advisers said he should send a note to the Democratic convention meeting in San Francisco announcing that he was not a candidate. But Wilson argued that the convention might deadlock and turn to him out of necessity. In any case, one doesn't refuse an offer that hasn't been made.

There were people crazy enough to encourage Wilson's dreams. One of these crazy people was the secretary of state, Bainbridge Colby, who at one point during the convention wired the

White House, "The outstanding characteristic of the convention is the unanimity and fervor of feeling for you. . . . I propose, unless otherwise definitely instructed, to take advantage of first moment to move suspension of rules and place your name in nomination."

The convention deadlocked, but Wilson's name was not put forth by Colby or anyone else. A small group of the President's friends had convinced the secretary of state that it was ridiculous to think even for a moment that Wilson could or should be elected to a third term. Finally, a compromise candidate, Governor James Cox, was chosen. At the White House an old man listened grimly to the news and then spewed forth a startling array of unbecoming and uncharacteristic obscenities. Woodrow Wilson had really hoped to become the first president to be elected to three terms.

SOURCE: Gene Smith, *When the Cheering Stopped* (New York: William Morrow, 1964), pp. 155–60.

ERA OF THE PUMPKIN COACH

★ ≋

"Let's keep what we've got: prosperity didn't just happen."
—HERBERT HOOVER CAMPAIGN SLOGAN IN 1928

★ SCRAPBOOK OF THE TIMES ★

- Six thousand corpses of American soldiers arrived from Europe on a single day in May 1921.
- In 1921 a bankrupt H. L. Hunt won his first oil well in a game of five-card stud in Arkansas.
- For no apparent reason Texas congressman Lindsay Blanton, a Presbyterian Sunday-school teacher and prohibitionist, inserted dirty words into the *Congressional Record* in 1921. He was censured by his colleagues, 293 to 0.
- Among the most popular books of 1924 was advertising executive Bruce Barton's *The Man Nobody Knows*, a book about Jesus as the quintessential businessman. Barton followed up this book with another a year later called *The Book Nobody Knows*, a work about the Bible.
- In 1924 the Sears catalogue carried for the last time an advertisement for "White Duck Emigrant Wagon Covers."
- Woodrow Wilson finally died in 1924, a year after his supposedly more vigorous successor, Warren Harding.
- Babe Ruth, who achieved fame first as a pitcher and only later as a hitter, earned a salary of $52,000 in 1925. Half a decade later his annual salary was up to $80,000.
- By answering an advertisement placed in a Tennessee newspaper by the American Civil Liberties Union in 1925, John T. Scopes became the center of a national controversy. The advertisement asked for a teacher who would volunteer to teach the theory of evolution in the public schools.
- The yo-yo was imported to America in the 1920s from the Philippines.

- By promising 50 percent interest on cash deposited for three months, Charles Ponzi, a lowly Boston clerk, quickly accumulated a fortune. For months he was able to operate successfully because practically no one withdrew their money. Finally the long lines outside his office shrank to nothing when he was proved to be a fraud. Ponzi then went to jail.

- Architects twice revised the blueprints of the Empire State Building to make the structure, originally planned for eighty stories, taller than the Chrysler Tower, which designers kept increasing in size in hopes of making it the highest skyscraper in the world. When finally completed, the Empire State Building reached 1,472 feet in the air, to a total of 102 stories.

- One of the pastimes of Zelda and Scott Fitzgerald in the 1920s was to ride on the rooftops of taxis in New York City. They also enjoyed bathing fully clothed in the fountains around the city.

- Scott Fitzgerald worried that his penis was small. One day he called Hemingway into a closet for an opinion. Hemingway reassured Scott that his penis was normal, and then took him on a tour of nude statues at a local museum to prove it.

- For appearing in a racy play called *Sex*, which she had written, Mae West, who was born with the improbable name Mary Jane, was sent to jail for ten days. Her only comment upon being released was that the prison underwear was awful.

- Walt Disney's first cartoon, *Plane Crazy*, appeared in 1928.

- Charles Loeb, an aspiring actor determined to become a movie star, mailed himself in 1929 from Chicago to a film studio in Culver City, California, in a box labeled, "Statue—handle with care." He arrived almost dead but told police he was happy because he had finally made it through the gates of a major studio.

- In 1929, Bing Crosby was arrested for drinking. When the judge asked him, "Don't you know about the prohibition law?" Crosby responded, "Nobody pays much attention to that." To which the judge replied, "Sixty days in jail. Next case."

- A major argument about fashion in 1939 was whether women tennis players should wear white stockings or nothing at all on their legs.

- The only nonwhite to be elected vice president of the United States was Charles Curtis, a Kaw Indian, who served under Herbert Hoover.
- In 1929 a government committee appointed in 1921 to study the problem of unemployment finally made its first report. Its conclusion? Unemployment was no longer a problem. The first chairman of the committee was Herbert Hoover.
- New York mayor Jimmy Walker once remarked: "A reformer is a guy who rides through a sewer in a glass-bottomed boat."

★ FDR's IRONIC CHOICE FOR PRESIDENT IN 1920 ★

At the end of World War I, Franklin Roosevelt launched a campaign to win the Democratic presidential nomination of 1920 for a friend. The friend was a prominent and popular member of the Wilson administration and seemed to Roosevelt to have a good chance of becoming president. Roosevelt himself was a rising star in the Democratic party. He carried a famous name, was heir to a modest fortune, and had distinguished himself during the ten short years he had been in politics. But he was only thirty-six years old, too young to be considered for the national ticket. So he put his talents to work for the benefit of a forty-five-year-old man with whom he had become friends. The two had met at the home of the secretary of the interior, where they frequently had Sunday dinners.

Unfortunately, Roosevelt was not sure if his friend was a Democrat or a Republican. At a dinner party one night he learned from the daughter of Senator Henry Cabot Lodge, Wilson's old foe, that his friend was a Republican.

This ended Roosevelt's efforts to make his friend president of the United States. The name of his friend was Herbert Hoover.

Footnote: The Republican Hoover did not find a place on the national ticket in 1920, but Franklin Roosevelt did, his youth notwithstanding. When Roosevelt was nominated for vice president by the Democrats, Hoover wrote him a warm letter, saying that FDR was a great public servant.

SOURCE: Gene Smith, *The Shattered Dream* (New York: William Morrow, 1970), pp. 85–86.

★ THE MAN WHO ALMOST BECAME ★
VICE PRESIDENT ★

"A legend about the Vice Presidential nomination that arose during the [1920 Republican] convention was for years afterward told by word of mouth and in print. It said that when Senator Knox of Pennsylvania was under consideration for the Presidential nomination, backers of him suggested an ingenious yet convincing pairing: Knox for President and Hiram Johnson for Vice President. The combination was geographically appealing. Pennsylvania on the Atlantic Coast, California on the Pacific—for that and other reasons it could have commanded, on paper, enough delegates to win. The proposers of the idea, hot with the fervor of invention, hurried to Johnson. As an allurement, they told him, in strict confidence, that Knox had heart disease and, if nominated and elected, would probably not live out the four-year term. Johnson blew up, emitted an indignant sentence: 'You would put a heart-beat between me and the White House!' He was, he told them sternly, a candidate for the Presidential nomination, and would take no less.

"Later, when the nomination of Harding was assured, his backers—so the legend said—seeking a running-mate, approached Johnson, as one who would help carry the West, and the progressive part of the Republican party. Again Johnson blew up. Again he declared that if he could not have first place, he would take nothing.

"Within four years, both Knox and Harding died. Had Johnson either made the arrangement that Knox's friends suggested, or consented to be the running-mate of Harding, he would have become President, would have stepped into the shoes that actually were filled by Calvin Coolidge."

SOURCE: Mark Sullivan, *Our Times* (New York: Scribner's, 1930–36), VI, 77n. Reprinted by permission of Charles Scribner's Sons.

★ HARDING COLLAGE ★

• Warren Harding's announcement that he had become engaged to Florence Kling DeWolf did not please everyone in Marion, Ohio, his hometown. When Florence's father heard of the engagement, he addressed the young suitor as "you God damned nigger," and promised personally to blow Harding's head off if he ever trespassed on the Kling premises.

- Mrs. Alice Roosevelt Longworth, daughter of T.R., once remarked that "Harding was not a bad man. He was just a slob."
- e. e. cummings described Harding in a poem as:

  The only man, woman or child
  who wrote a simple declarative sentence
  with seven grammatical errors.

- Harding was once heard to remark: "Oftentimes . . . I don't seem to grasp that I am President."
- Harding accomplished little during his term of office, but he did have one claim to fame: he was the first president to know how to drive an automobile.
- Norman Thomas was the leader of the American Socialist party and six times a candidate for president. He also happened to be a friend of Warren Harding, having worked for Harding as a newsboy on the Marion *Star.*
- Harding's sudden death shocked the nation but not Dr. Emmanuel Libman, a heart specialist. Eight months before Harding died, Libman observed the President at a party and predicted to a friend that Harding would suffer a fatal coronary within six months.

★ HARDING AND THE PRESS CONFERENCE ★

Every president strives for good relations with the press. Warren G. Harding, a newspaper publisher for thirty-five years before coming to Washington, was no exception. Harding's plan was simple. As a former newspaperman, the new president thought he could speak to reporters in their own lingo, frankly and openly. Press conferences were to have a casual, off-the-cuff atmosphere, in which a correspondent could ask the chief executive virtually any question on any subject. The President would then respond in a free and easy manner.

In one of his first press encounters, Harding discovered that government policies were very complex, especially without the benefit of briefings on special questions. The Washington Conference on Naval Disarmament was then in session. Eventually, the conference would hammer out the Four Power Treaty which would set ratios on the possession and production of large warships among the world's seagoing powers. A correspondent asked Harding if Japan, for the purposes of the ratios, was considered a Pacific island or a part of the Asian mainland. Harding did not know, but rather than appear ignorant on a subject of current importance, he guessed.

Unfortunately, he guessed wrongly. Secretary of State Charles Evans Hughes corrected the embarrassing presidential blunder as diplomatically as possible, while Harding himself accepted full blame for his political faux pas. Hughes made one strong suggestion—that the President abandon his informal approach to the press. From that time on, all questions were required in writing in advance. Harding's press relations, even with the new rule, were superb. It was perhaps the most successful aspect of his administration.

SOURCE: James E. Pollard, *The Presidents and the Press* (New York: Macmillan, 1947), p. 705.

★ A SILK PURSE FROM A SOW'S EAR ★

The proverb says, "You can't make a silk purse out of a sow's ear," but in 1921 the founder of one of America's leading research companies proved that you can. Arthur D. Little began by ordering from a Chicago meat-packing company ten pounds of gelatin "manufactured wholly from sow's ears." Then, using synthetic processes, he spun the gelatin into an artificial silk thread. From this he had a purse woven, "of the sort which ladies of great estate carried in medieval days— their gold coin in one end and their silver coin in the other. It is one of which both Her Serene and Royal Highness the Queen of the Burgundians in her palace, and the lowly Sukie in her sty, might well have been proud." Little put the purse on public display and issued a pamphlet describing how the miracle was accomplished.

SOURCE: Daniel Boorstin, *The Americans: The Democratic Experience* (New York: Random House, 1973), pp. 545–46.

★ CIGARETTE SMOKING IS GOOD FOR ★ YOUR HEALTH

An advertisement for Lucky Strikes used in the 1920s:

"Instead of eating between meals . . . instead of fattening sweets . . . beautiful women keep youthful slenderness these days by smoking *Luckies*. . . . Lucky Strike is a delightful blend of the world's finest tobaccos. These tobaccos are toasted—a costly process which develops and improves the flavor. That's why *Luckies* are a delightful alternative for fattening sweets. *That's why there's real health in Lucky Strikes.* For years this has been no secret to those who keep fit and trim. They know that *Luckies* steady their nerves and do not hurt their physical condition. They know that *Lucky Strikes* are the favorite cigarette of many prominent athletes who must keep in good shape. They respect the opinions of 20,679 physicians who maintain that *Luckies* are less irritating to the throat than other cigarettes."

SOURCE: James Wood, *The Story of Advertising* (New York: Ronald Press, 1958), p. 378.

★ NEW YORK PRESS UNITES ★

The newspaper business in the early part of the twentieth century was fiercely competitive. But in September 1923 a printers' strike in New York City caused the morning newspapers to unite. The union of the papers resulted in one of the strangest and most impressive mastheads of all time:

SOURCE: Mark Sullivan, *Our Times* (New York: Scribner's, 1930–36), VI, 602. Reprinted by permission of Charles Scribner's Sons.

★ QUIPS FROM H. L. MENCKEN ★

- "Men have a much better time of it than women. For one thing, they marry later. For another thing, they die earlier."
- "Say what you will about the Ten Commandments, you must always come back to the pleasant fact that there are only ten of them."
- "Conscience: the inner voice which warns us that someone may be looking."

★ COOLIDGE COLLAGE ★

- In 1924, Coolidge's son died after getting a blister on his toe while playing lawn tennis.
- Coolidge often displayed sadistic tendencies. Once he asked a Secret Service agent to bait his fishhook. Just as the man was hooking the worm, Coolidge jerked the line and drew blood from the agent's fingers.
- Coolidge loved having his head rubbed with Vaseline while he ate breakfast in bed.
- Coolidge was an exceedingly stingy man. At the White House he demanded change from servants whom he had given money to for newspapers. When the servants kept the change, Coolidge would wander around the mansion saying, "Somebody owes me seven cents."
- When Hoover was elected president, Coolidge gave him just one bit of advice: "If you don't say anything, you won't be called on to repeat it."
- Coolidge's modest ways were revealed by his choice of residence. Both before and after his term of office, he lived in Northampton, Massachusetts, in one half of a rented two-family house.
- After Coolidge left the White House, he became a daily columnist, receiving one dollar for every word he wrote, or about $200,000 a year—over three times his salary as president. His skill as a columnist was revealed in his famous

explanation of unemployment: "When more and more people are thrown out of work, unemployment results."

- When Coolidge died, Dorothy Parker asked, "How can they tell?"

★ COOLIDGE WON'T GO TO HELL ★

"A [Massachusetts state] legislative session was nearing its close, and, as was the usual practice, the two leaders [the Speaker of the House of Representatives, and Calvin Coolidge, the president of the Senate] were conferring to determine what matters would be given precedence for action prior to the adjournment. Present also was a Senator from Boston who was vigorously determined that a bill that directly affected his district should be taken up immediately and not be carried over at a later date. In opposition to this insistence, Mr. Coolidge kept saying, 'Senator, I don't think that's important enough in this rush hour.'—'Senator, I don't think we can do it.'

"Finally, indignant and angry, the legislator snapped at the Senate's presiding officer, 'You can go to hell.'

"'Senator,' Mr. Coolidge shot back, 'I've looked up the law and I find I don't have to.'"

SOURCE: Edward Lathem, *Meet Calvin Coolidge* (Brattleboro, Vt.: Stephen Greene Press, 1960), p. 149.

★ CALVIN COOLIDGE A WIT IN HIS OWN WAY ★

Calvin Coolidge possessed the wry sense of humor characteristic of his native Vermont. Examples from his press conferences:

President: I haven't had any report from the Tariff Commission on butter, or straw hats, or gold leaf. I have a report on cotton gloves.

President: I think the press already knows that I am expecting to attend the Fair—tomorrow, isn't it, Mr. Sanders?

Mr. Sanders: Yes, tomorrow afternoon about 2:00 o'clock.

Reporter: It isn't likely you will say anything tomorrow at the Fair?

President: No. I am just going as an exhibit.

SOURCE: Howard Quint and Robert Ferrell, eds., *The Talkative President* (Amherst: University of Massachusetts Press, 1964), pp. 18, 19.

★ COOLIDGE PUSHES THE BUTTON ★

President Coolidge was a practical joker. One prank grew out of his discovery, after an early-morning walk, of an alarm button on the front porch of the White House. "Feigning he was tired," his Secret Service agent recalled, "he leaned against the button and pressed it. His solemn, immobile expression unchanged, he walked hurriedly into the house and from behind the safety of the living-room curtains peeked out and saw two policemen come tearing across the lawn, survey the scene, and, finding no one, return to the guard house. He pushed the button two more times and each time he would, without change of expression, watch the excitement that resulted."

SOURCE: Edward Latham, *Meet Calvin Coolidge* (Brattleboro, Vt.: Stephen Greene Press, 1960), p. 158.

★ COOLIDGE PUTS HAIR ON JOHN ADAMS'S HEAD ★

Calvin Coolidge did not like the oil painting of John Adams that hung in the Red Room, which he saw frequently from his table in the State Dining Room. He particularly disliked Adams's bald head. But instead of having the portrait removed, he had the chief usher of the White House, Ike Hoover, put some hair on Adams's head. Hoover obeyed and had an artist smear some turpentine on the bald pate to take off the shine and give it the appearance of a little hair. Coolidge later remarked to Hoover that Adams seemed to have grown some hair.

SOURCE: Irwin Hood Hoover, *Forty-two Years in the White House* (Boston: Riverside Press, 1934), pp. 128–29.

★ COOLIDGE WITHOUT RECOURSE ★

A story about Silent Cal by William V. Hodges, treasurer of the Republican party during Coolidge's administration:

"I was lunching with President Coolidge one day when we were joined by an author who, of his volition and without approach to Mr. Coolidge, had written and published a biography of Calvin Coolidge. He wanted to present a copy of his book to the President. Mr. Coolidge was gracious enough to accept the book. Taking another copy of his work from under his arm, the author told the President that he would deem it a great honor if the President would write his name in the volume. The President did so. This is what he wrote: 'Without recourse, Calvin Coolidge.'"

SOURCE: Richard Peete, *Anecdotes of the Jealous Mistress* (Denver: University of Colorado Law Review, 1959), p. 4.

★ WHY CALVIN COOLIDGE CHOSE NOT TO RUN ★

One of the most famous and mysterious announcements ever made by a president was Calvin Coolidge's 1927 statement, "I do not choose to run for president in 1928." The statement was characteristically terse, but seemingly unclear. Coolidge had been expected to run again, and in the absence of an explanation his declaration seemed suspicious. Did he choose not to run because he wanted to be drafted? Pundits went wild with speculation and examined closely the import of every word.

They need not have bothered. The announcement meant exactly what it seemed to mean. Why then did Coolidge not explain his decision and end the dark guesses about his motives? He didn't want to reveal that he was physically unable to serve another term. A short time before, he had suffered a heart attack. Not until recently was this known.

SOURCE: "The Reminiscences of Claude Moore Fuess" (1962), pp. 78–79 in the Oral History Collection of Columbia University.

★ MORE ON COOLIDGE ★

Will Rogers, the famous humorist, once asked Calvin Coolidge how he kept fit in a job that had broken the health of Woodrow Wilson. "By avoiding the big problems," Coolidge replied in all seriousness.

The president further eased the burdens of his office by confining himself to four hours of work a day and by taking a nap every afternoon.

SOURCE: Isabel Leighton, ed., *The Aspirin Age* (New York: Simon & Schuster, 1949), p. 145.

★ AL SMITH HOPES HE'S RIGHT ★

"It seems that one day in Albany, in the midst of an important political convention, but not forgetful of high jinks, Herbert Lehman, Al Smith, Jimmy Walker, and many others of the old timers, had a big night of it. The next day was a Catholic holy day. Al Smith and Walker and the other Catholic members of the group decided that they had to make the early mass. They all occupied the same suite of rooms. As they tiptoed, groggy-eyed, through the rooms and looked at Lehman and others of the Jewish faith sleeping so quietly and beautifully, Al Smith turned to Walker and said, 'Gee, I hope we're right!'"

SOURCE: B. A. Botkin, ed., *A Treasury of American Anecdotes* (New York: Random House, 1957), p. 41.

★ BRINKLEY AND THE BILLY GOATS ★

Dr. John R. Brinkley was no ordinary doctor. Since around 1913 he had considered the possibility of performing a sex-gland transplant operation to restore lost potency. It had been done successfully in Europe by transplanting chimpanzee glands in human males. Patients reportedly left the operating table with potency assured. After Brinkley settled in Milford, Kansas, in 1918, he decided to attempt an operation. Instead of chimpanzee glands, however, Dr. Brinkley substituted billy goat's glands. His first patient was a middle-aged man who had long desired a family. Both doctor and patient knew the operation was an experiment, but they agreed to take the gamble. Some time later the man presented Brinkley with "a small roll of bills." The man's wife had had a son. Brinkley then operated on other patients.

A prominent California newspaperman, following his successful transplant, suggested that Brinkley raise his fee from $50 to at least $750. Fifty dollars was simply too small a sum to sound impressive. Soon Dr. Brinkley had paid off the debt on his new Milford hospital. By 1928 his fame had spread far and wide. Thirty to forty young, vigorous billy goats arrived daily from Arkansas to supply the needed number of sex glands. Willing patients were even easier to obtain. The doctor examined forty each day. Brinkley's income was estimated at anywhere between $50,000 a year and $30,000 a week. He owned three airplanes and had opened his own bank. Brinkley was also the founder-owner of the first radio station in Kansas.

But 1928 turned out to be the beginning of hard times for Brinkley. With an article in the journal of the American Medical Association entitled "John R. Brinkley—Quack," Morris Fishbein and the AMA launched a close-down-Brinkley campaign. There was no medical evidence, they claimed, that substituting billy goat glands for human glands would improve anyone's potency. The state Board of Medical Examination and Registration soon rescinded Brinkley's license. The courts upheld the board's action. To worsen matters, Brinkley's radio broadcasting permit was about to be revoked. It looked as though Brinkley's whole world would cave in.

To save his business and rescue his reputation, Brinkley entered politics. He announced as an independent candidate for governor of Kansas. Unfortunately, the ballots for the 1930 election were already printed, forcing Brinkley to launch a write-in campaign. The state attorney general, after much political haggling, ruled that only ballots marked "J. R. Brinkley" and followed by the proper "X" would be counted. Ballots not properly marked—ballots cast for "Dr. Brinkley," "Brinkley," or with the name misspelled—would all be discarded. On election day literally thousands and thousands of Brinkley write-in ballots were invalidated on the flimsiest pretexts. The doctor lost the tight three-way election. Thirty years later both the Republican and Democratic candidates admitted that Brinkley would probably have won had he received every vote cast for him. Brinkley ran for governor again in 1932 and 1934, both times filing early enough to receive a proper place on the ballot. He was defeated in each race, however, by Republican Alf M. Landon.

The 1930 governor's race did not ruin Brinkley financially. He opened a new radio station just across the Rio Grande in Mexico—free of Federal Communications Commission regulations—and cranked out 50,000 watts, broadcasting to almost the entire Middle West. Dr. Brinkley did, however, retire from medical practice.

SOURCE: Francis Schruben, *Kansas in Turmoil* (Columbia, Mo.: University of Missouri Press, 1969), pp. 28–29.

★ THE NOT-SO-ROARING TWENTIES ★

According to figures on the 1920s compiled by the Brookings Institution, the decade's reputed bubble of prosperity was never as big as it seemed. A handful of people did indeed make fortunes in the stock market and in real estate, but most people were not so lucky. In 1929 only 2.3 percent of American families enjoyed an income in excess of $10,000 a year. Just 8 percent had incomes greater than $5,000. Of the others, 71 percent lived on incomes below $2,500; 60 percent on incomes below $2,000. And over 21 percent had incomes of less than $1,000.

The Brookings study established that families living on less than $2,000 did not have enough money to meet the bare necessities of life. Which means that in the golden year of 1929, 60 percent of American families were living in poverty, and another 10 percent were close to it.

SOURCE: Frederick Lewis Allen, *The Big Change* (New York: Harper & Brothers, 1952), p. 144.

★ THE PRIDE OF A GANGSTER ★

Joe Saltis, a bootlegger from Chicago's South Side, had a special dream. Like most gangsters from the 1920s era, Saltis had a large ego. When he retired from the rackets in 1930, he planned to feed that ego.

Saltis moved to a farm at Barker Lake, Wisconsin, a luxury accommodation complete with a nine-hole golf course, clubhouse, ponies, deer, and good fishing grounds. The small town nearby had sixty-two voters, twenty-six of whom worked for Saltis. But that was not enough for a majority at the polls, so the former gangster hired five more people and changed the name of the town to Saltisville.

"What I want is for my kids to be able to look in the United States Postal Guide and see their town, Saltisville," claimed the unabashed bootlegger. A check with the Post Office, however, reveals that the town of Saltisville no longer exists.

SOURCE: Jay Robert Nash, *Bloodletters and Badmen* (New York: M. Evans and Company, 1973), p. 484.

WAR ON DEPRESSION, WAR ON EUROPE

★ ≋

"These really are good times, but only a few know it."
—HENRY FORD, MARCH 15, 1931

★ SCRAPBOOK OF THE TIMES ★

- When Herbert Hoover invited the Negro wife of a congressman to the White House for tea, he was officially denounced by the state legislature of Texas.
- Just a month before the stock market crashed in October 1929, the vice-chairman of General Motors wrote an article for the *Ladies Home Journal* entitled "Everybody Ought to Be Rich."
- The Democratic theme song, "Happy Days Are Here Again," comes from a movie called *Chasing Rainbows,* which opened to universal derision two months *after* the Great Crash.
- During the 1930s, Bernard E. Smith made a fortune in the stock market by following the rule of thumb that the market would decline every time Hoover issued an optimistic statement about recovery.
- A common sign held up by hitchhikers during the fall of 1932 read: "If you don't give me a ride, I'll vote for Hoover."
- In the midst of the 1932 campaign, when it appeared that Hoover would lose badly to FDR, the Republican vice presidential nominee blurted out at one stop that the problem with the country was that the average voter was "too damn dumb" to understand the administration's policies.
- Despite his radical politics, John L. Lewis, head of the powerful United Mine Workers, voted for Hoover over FDR in 1932. Lewis voted Republican his whole life except in the election of 1936, when he cast a ballot for Roosevelt.

- When one of the Du Ponts was advised by an advertising agency in the middle of the depression to sponsor a radio program on Sunday afternoon, he remarked, "At three o'clock on Sunday afternoon everybody is playing polo."

- Eleanor Roosevelt did not like to smoke, but she often smoked at White House dinners as part of her campaign to "help" women.

- In 1933, U.S. customs prohibited the importation of a book which contained a picture of Michelangelo's *Last Judgment*. The ban was lifted when publicized in the newspapers.

- In the 1930s, Mae West's name was never allowed to be mentioned in the Hearst chain of newspapers.

- During the 1936 campaign William Randolph Hearst advised the Republican party that presidential candidate Alf Landon should be encouraged to make as few speeches as possible.

- Encouraged by the success of the Veterans of Foreign Wars, which in 1936 persuaded Congress to give a bonus to the veterans of World War I, students at Princeton University established an organization to demand compensation for another group of veterans: the Veterans of Future Wars. The students argued that the government should immediately give every male citizen between the ages of eighteen and thirty-six a bonus of $1,000 each. Then, when the next war came along, every soldier would know that dying would not cheat him out of an extra paycheck. Within ten days, more than 120 branches of the organization were established at other colleges. Female college students founded their own group: the Future Gold Star Mothers.

- To give businessmen in 1939 a longer Christmas season, FDR ordered that Thanksgiving be celebrated one week earlier than usual.

- While running for president in 1940, Wendell Willkie used the slogan: "Roosevelt for ex-President."

- During the campaign of 1940, *Fortune* magazine reported that only 7.3 percent of American women and only 5.8 percent of American men believed that Eleanor Roosevelt should remain active in politics if her husband lost the election.

- The first freeway in America was opened on December 30, 1940, in Los Angeles.

★ THE INDIAN WHO BECAME VICE PRESIDENT ★

He is almost forgotten now, but America once elected a part Indian as vice president, literally a heartbeat away from the presidency. Charles Curtis was a U.S. senator from Kansas, Senate majority leader, and vice president of the United States under Herbert Hoover.

Curtis was born in 1860 in Kansas. His mother died in 1863 and his father was dishonorably discharged from the Union Army and served time in the Missouri State Penitentiary. By 1866, Charley Curtis was being raised by his maternal grandparents on the Kaw Indian reservation in Council Grove, Kansas. Curtis's grandmother, Julie Gonville Pappan, was the granddaughter of White Plume, the Kansa-Kaw chief who had helped Lewis and Clark in 1804. White Plume's daughter had married a French-Canadian fur trapper, Louis Gonville, and their daughter Julie had married another French-Canadian fur trapper, Louis Pappan. Reflecting his mother's and grandparents' heritage, Curtis grew up speaking Kaw and French before he learned English.

Because he spoke Kaw, Curtis fit in easily on the reservation. He was a good horse rider and shot bows and arrows. In those frontier days, the Kaw reservation was frequently raided by still nomadic Cheyenne Indians. On one such attack, the horses were hidden and Curtis volunteered to make the sixty-mile trip from Council Grove to Topeka on foot to get help. This "cross-country run" made Curtis a celebrity in Topeka, but convinced his paternal grandparents that he should be raised in the more "civilized" atmosphere of Topeka. Curtis lived the next few years of his life in Topeka, where he often jockeyed horses in the local races.

In 1873, Curtis's paternal grandfather died. Curtis left Topeka and wanted to join his maternal grandparents, who were moving with the Kaw tribe to the new reservation in the Indian Territory of present-day Oklahoma. His grandmother Julie Pappan, however, talked him out of rejoining the tribe. While she would have loved to have him live with her, she told her grandson that if he lived on the reservation, he would end up without an education or future prospects. The next morning, as the Kaw wagons pulled out and headed south to what would become Oklahoma, young Charles Curtis mounted his pony and returned to Topeka. He had irrevocably entered the white man's world. "It was the turning point in my life," he later recounted.

Back in Topeka, Curtis lived with his paternal grandmother, a woman who "regarded being both a Methodist and a Republican as essential for anyone who expected to go to heaven." Curtis graduated from high school, studied law, and was admitted to the Kansas bar at age 21. At age 24, he married and won his first political race, as Shawnee County attorney.

Curtis had started on a long and very successful political career. In 1892, he was elected to the U.S. House of Representatives as a Republican, in a year when the state of Kansas voted for the Populist candidate for president and elected a Populist governor. In 1907 he was elected to the U.S. Senate, and in 1925 became Senate majority leader.

In 1924, Curtis was widely discussed as a vice presidential candidate for Calvin Coolidge. His wife, however, was seriously ill at the time, and Curtis did not attend the Republican convention. He told his sister, "I would not leave [my wife] Anna now to become president of the United States, and certainly not for the vice presidency." His wife died later that year.

In 1928, Curtis announced his candidacy for president, but ran an ineffectual campaign. Herbert Hoover was nominated on the first ballot. To balance the ticket, Republicans needed a farm state man and Curtis was it. That November, America elected as vice president the man who had grown up on the Kaw reservation.

Hoover and Curtis had never been political allies, and their distrust of each other lasted throughout Hoover's term. Curtis was never an inside player in the administration. By 1932, with the onset of the Great Depression, Herbert Hoover and Charles Curtis met defeat at the polls. Curtis's political career was over. He died four years later.

Source: Mark O. Hatfield, *Vice Presidents of the United States, 1789–1993* (Washington: U.S. Government Printing Office, 1997), pp. 373–81.

★ HOOVER COLLAGE ★

- While a student at Stanford, Hoover opposed elitist fraternities and founded his own group, the Barbarians.
- At Hoover's inauguration in 1929, Chief Justice William Howard Taft bungled the swearing-in ceremony. Instead of asking Hoover to "preserve, protect, and defend the Constitution of the United States," he used the words, "preserve, maintain, and protect." The slip caused low murmurs among the spectators.
- Hoover was the first president to have a telephone right on his desk. All previous presidents kept the instrument in an adjoining room.
- While an engineer, Hoover translated a sixteenth-century work on mining written in Latin.
- Hoover required White House servants to be "invisible." Whenever he or the first lady appeared, the servants would jump into the nearest closet to avoid being seen.
- After learning that Maine had gone Democratic in the September election in 1932, Hoover ordered an aide to determine the truth of the old saying, "As Maine goes, so goes the nation."
- In his memoirs Hoover asserted that people did not turn to apple-selling in the depression because they were unemployed, but because apple-selling was a highly profitable business. "Many persons," he wrote, "left their jobs for the more profitable one of selling apples."

★ AN EMPTY POT AND A VACANT GARAGE ★

With the coming of the depression, the Republican slogan of 1928, "A chicken in every pot and two cars in every garage," seemed a hideously bad joke—especially to the reporter who coined the phrase. By 1933 the author of the slogan was out of work and driven to begging to keep his wife and children from starving.

SOURCE: Olive Clapper, *Washington Tapestry* (New York: Whittlesey House, 1946), p. 10.

★ MICKEY MOUSE BANNED ★

The banning of a Mickey Mouse cartoon is the last thing one would expect, but in 1932 it happened. The cartoon showed a cow in a pasture reading a book that was considered obscene. The book was Elinor Glyn's *Three Weeks*, which told the story of a young Englishman's affair with a Russian queen. The book had been banned in Boston in 1908 and was considered vulgar and boring by moralists and critics alike. It is a mystery how the book came to be put in a Walt Disney cartoon.

SOURCE: Anne Lyon Haight, *Banned Books*, 3d ed. (New York: R. R. Bowker, 1970), pp. 67, 92.

★ HUEY LONG'S HORSE ★

T. Harry Williams opens his biography of Huey Long with this anecdote:

"The story seems too good to be true—but people who should know swear that it is true. The first time that Huey P. Long campaigned in rural, Latin, Catholic south Louisiana, the local boss who had him in charge said at the beginning of the tour: 'Huey, you ought to remember one thing in your speeches today. You're from north Louisiana, but now you're in south Louisiana. And we got a lot of Catholic voters down here.' 'I know,' Huey answered. And throughout the day in every small town Long would begin by saying: 'When I was a boy, I would get up at six o'clock in the morning on Sunday, and I would hitch our old horse up to the buggy and I would take my Catholic grandparents to mass. I would bring them home, and at ten o'clock I would hitch the old horse up again, and I would take my Baptist grandparents to church.' The effect of the anecdote on the audiences was obvious, and on the way back to Baton Rouge that night the local leader said admiringly: 'Why, Huey, you've been holding out on us. I didn't know you had any Catholic grandparents.' 'Don't be a damn fool,' replied Huey. 'We didn't even have a horse.'"

SOURCE: T. Harry Williams, *Huey Long* (New York: Knopf, 1970), p. 3. Reprinted by permission of Alfred A. Knopf, Inc.

★ FDR Collage ★

- While president, Franklin Roosevelt always slept with a gun under his pillow. Eleanor usually carried a pistol in her pocket-book and glove compartment. White House servants called the presidential weapons "His" and "Hers."
- Because he was handicapped by polio, FDR's greatest fear was fire. He never liked being left alone in a room with a fire burning in a fireplace.
- One of Roosevelt's favorite stories concerned the wife of a foreign diplomat who lost her panties as she was going up to shake the President's hand during a state dinner.

★ FDR's Impressive Lineage ★

Genealogists have determined that FDR was related to the following eleven presidents:

George Washington

John Adams

James Madison

John Quincy Adams

Martin Van Buren

William Henry Harrison

Zachary Taylor

U. S. Grant

Benjamin Harrison

Theodore Roosevelt

William Howard Taft

Source: Joseph Kane, *Facts About the Presidents*, 2d ed. (New York: H. W. Wilson, 1968), p. 223.

★ FDR a Greater Man Than Christ ★

"A hillman told me the other day that Roosevelt was the greatest man that ever lived. 'Greater'n Jesus Christ,' he said solemnly. 'Christ said, "Follow me and ye shall not want." Roosevelt says, "Set down, boys, and I'll bring hit to ye!"'"

Source: *Esquire*, April 1937, p. 95.

★ Letter from a Jackass ★

California congressman John Steven McGroarty, responding to a letter from a constituent in 1934: "One of the countless drawbacks of being in Congress is that I am compelled to receive impertinent letters from a jackass like you in which you say I promised to have the Sierra Madre mountains reforested and I have been in Congress two months and haven't done it. Will you please take two running jumps and go to hell."

Source: John F. Kennedy, *Profiles in Courage* (1956; rpt. New York: Perennial Library, 1964), p. 9.

★ The Word That Does Not Exist ★

One day a tired linguist working on the highly esteemed second edition of the 1934 Merriam-Webster New International Dictionary carelessly placed a slip of paper containing the abbreviation for the word "density" ("D. or d. Density") on the pile of slips for words beginning with the letter *d*. Then another linguist, thinking the slip was in the right place but that the entry on it had been wrongly punctuated, pushed the first four letters together to form the word "dord." He thoughtfully added the descriptive letter *n* for "noun." Clearly "dord" was a noun—it rhymed with "board," "cord," and "lord."

The astute editors of the dictionary soon discovered the errant entry but decided to play a joke on the public and leave it in. They wanted to find out if anyone would catch the "mistake." So, appearing in the August second edition of the 1934 Merriam-Webster dictionary, on page 771, in the right-hand column, sandwiched in between the words "Dorcopsis" and "doré," is the following entry: "dord (dôrd), *n. Physics & Chem.* Density."

"Dord" stayed in the dictionary through several printings, but was finally dropped when new editors took over.

SOURCE: William Morris and Mary Morris, *Harper Dictionary of Contemporary Usage* (New York: Harper & Row, 1975), pp. 276–77.

Dor·cop′sis (dôr·kŏp′sĭs), *n.* [NL., fr. Gr. *dorkas* gazelle + *-opsis*.] *Zool.* A genus of small kangaroos of Papua.
dord (dôrd), *n. Physics & Chem.* Density.
‖**do′ré′** (dô′rā′), *adj.* [F.] **a** Golden in color. **b** *Metal.* Containing gold; as, *doré* silver. — *n.* = DORÉ BULLION.
‖**do′ré′** (dô′rā′), *n.* [F., gilded. See 2d DORY.] The wall-eyed pike. *Fr. Canadian.*

The errant entry. (*Webster's New International Dictionary*, 2d ed., 1934.)

★ MIXED METAPHORS ★

Republican campaign manager John Daniel Miller Hamilton, in July 1936, responding to charges that the Republicans were resorting to anti-Semitism: "There is not an iota of truth in such a thing, and it is a deliberate attempt by those other people to throw a dust cloud when they know their ship is sinking. We have a red herring in every campaign, and apparently this is the first such attempt."

SOURCE: Ernest Sutherland Bates and Alan Williams, *American Hurly-Burly* (New York: Robert M. McBride, 1937), p. 193.

★ FDR CLEARS HIS CONSCIENCE ★

Politicians sometimes find it difficult to keep their promises. Even while campaigning for president in 1932, Franklin Roosevelt suspected that one thing a Roosevelt administration would not offer the American people was a cut in the federal budget. True, one of the major planks in the Democratic platform that year was a reduction in government expenditures. But Roosevelt had been generally silent on that particular plank throughout his campaign. Except once. On October 19, speaking in Pittsburgh, FDR promised that if elected he would reduce the government budget.

By the next presidential election, in 1936, the federal budget was more unbalanced than it had ever been in American history. New Deal expenditures were still rising. And Republicans frequently quoted the 1932 Pittsburgh speech, accusing the President of deceiving the people.

But Roosevelt would outdo them. The first major speech of the 1936 race would be made in Pittsburgh, at the same ball park as the earlier address. The President called in Samuel Rosenman, a close friend and adviser, and asked him to carefully go over the 1932 speech and prepare a draft explaining the apparent discrepancies between his promises and his performance. Rosenman returned to the White House that night and claimed that he had the explanation. The President was delighted. "Fine, what sort of explanation would you make?" he asked.

"Mr. President," replied Rosenman, "the only thing you can say about that 1932 speech is to deny categorically that you ever made it."

SOURCE: Samuel Rosenman, *Working with Roosevelt* (New York: Harper & Brothers, 1952), pp. 86–87.

★ IRONIC ORIGINS OF FDR's ★
COURT-PACKING PLAN

Voices of outrage greeted Franklin Roosevelt's 1937 Supreme Court packing plan. Republicans declared that their worst suspicions about the President had been proven true; liberals argued that the plan threatened the constitutional balance of powers, and even some ardent New Dealers conceded that the patrician from Hyde Park had gone too far. To many Americans, Roosevelt appeared ready to turn the United States into a dictatorship.

The irony of all this was that the plan had been based almost completely on suggestions made in 1913 by one of the very members of the Supreme Court whom Roosevelt wanted to force off the bench—James McReynolds, attorney general under Woodrow Wilson and now a leading foe of the New Deal. McReynolds had written that judges older than seventy ought to be required to retire, since some had "remained upon the bench long beyond the time that they are able to adequately discharge their duties, and in consequence the administration of justice has suffered." FDR agreed entirely, but there was one significant difference between his plan and McReynolds's. The future justice had specifically exempted members of the Supreme Court from the age limit.

SOURCE: William E. Leuchtenburg, "The Origins of Franklin D. Roosevelt's 'Court-Packing' Plan," *Supreme Court Review* (Chicago: University of Chicago Press, 1966), pp. 387–92.

★ AND HE SOLD A LITTLE FLOUR ON THE SIDE ★

The depression produced some strange politicians, and W. Lee "Pappy" O'Daniel, governor of Texas, was one of them. O'Daniel was born in Ohio, lived in Kansas, and in 1925 moved to Fort Worth, Texas, as sales manager of a flour milling company. In 1927, Pappy began a statewide radio program, where he, backed by a three-man group of hillbilly musicians known as the Light Crust Doughboys, sang country and religious tunes, gave advice to housewives, recited poetry, advocated thrift and morality, and sold a little flour. Each day at noon, the call words "Please pass the biscuits, Pappy" announced the beginning of the broadcast. By 1938, O'Daniel was president of his own flour company and a statewide personality. Hillbilly Flour, his firm's product, sold well throughout the area.

With the 1938 race for the governorship approaching and the usual crop of political candidates filing, a listener wrote in and told Pappy to have a go at the office. On his next show O'Daniel read the letter and requested his audience's advice. In one week Pappy received 54,449 "you should run" letters from all across the state. Pappy and twelve other candidates filed in the Democratic primary.

At first newspapers and the other contenders treated O'Daniel's candidacy as a joke. Pappy had no headquarters, no campaign manager, and no practical knowledge of politics. By his own admission, he had neither cast a vote nor paid a poll tax his whole life. His announced platform was the Ten Commandments and the Golden Rule. "I don't know whether I'll get elected," admitted Pappy, "but, boy, it sure is good for the flour business."

Slowly, the political establishment realized that O'Daniel's candidacy was very real and that the crowds flocking to his campaign rallies had not come simply to hear a good band play for free. On election day all these people who had been dropping nickels and dimes into Pappy's special-donation flour barrels turned out at the polls, and O'Daniel swept his twelve opponents with over 50 percent of the vote. In November, Democrat O'Daniel leveled his Republican challenger, 473,526 to 10,940.

In office, Pappy's campaign promises, particularly on old-age benefits, were watered down by a dubious legislature, but Pappy's radio show continued full steam. He was reelected governor two years later, more on account of his continuing broadcasts than his record at the statehouse. O'Daniel later became a U.S. senator from Texas, narrowly defeating a young Lyndon B. Johnson in the 1941 Democratic primary. In 1956 and again in 1958, Pappy ran for governor and polled a respectable percentage of the votes, but both times lost.

SOURCE: Seymour V. Connor, *Texas: A History* (New York: Crowell, 1971), p. 342.

★ SURPRISE AT PEARL HARBOR ★

It is almost too incredible to be true. Beginning in 1931, ten years before the Japanese surprise attack on Pearl Harbor, every graduate of the Japanese Naval Academy had to answer the following question as part of his final examination: "How would you carry out a surprise attack on Pearl Harbor?" The question remained on the cadets' exam every year until the beginning of the war in the Pacific. It is not known if the Japanese high command used any of the answers from the ten-year period while planning the real attack.

Source: Quincy Howe, *Ashes of Victory: World War II and Its Aftermath* (New York: Simon & Schuster, 1972), p. 161.

★ Eleanor Shakes a Hand ★

Eleanor Roosevelt was a special first lady. Her vibrancy and energy were felt throughout the country during and after FDR's long administration. When World War II was raging, Eleanor visited American servicemen worldwide, often returning with personal messages for their families. But in one particular South Pacific hospital her enthusiasm led to embarrassment. Suddenly, to the horror of her escorts, the first lady burst unannounced into a particular ward and started handshaking and kissing the wounded. The ward, her escorts had not been able to warn her, was reserved for American soldiers "wounded" by venereal disease.

Source: *Time*, November 16, 1962, p. 29.

★ Never Kill a Republican ★

A story told by FDR to his cabinet:

"An American Marine, ordered home from Guadalcanal, was disconsolate and downhearted because he hadn't killed even one Jap. He stated his case to his superior officer, who said: 'Go up on that hill over there and shout: "To hell with Emperor Hirohito!" That will bring the Japs out of hiding.'

"The Marine did as he was bidden. Immediately a Jap soldier came out of the jungle, shouting: 'To hell with Roosevelt!'

"'And of course,' said the Marine, 'I could not kill a Republican.'"

Source: William D. Hassett, *Off the Record with FDR*. (New Brunswick, N.J.: Rutgers University Press, 1958), p. 175.

★ SIR WINSTON—PART INDIAN? ★

Winston Churchill, one of Great Britain's greatest prime ministers, was a direct descendant of Iroquois Indians. Jennie Jerome, Winston's mother, was born in Brooklyn, New York, in 1850. Her mother was one-quarter Iroquois. Jennie married Lord Randolph Churchill, making their son one-sixteenth Iroquois. While being part Indian was somewhat of an embarrassment to the Jerome family, Winston himself was delighted to be a descendant of American Indians.

SOURCE: William Manchester, *The Last Lion: Winston Spencer Churchill: Visions of Glory* (New York: Dell, 1983) p. 101.

★ FDR AND THE CASE OF THE MISSING MAP ★

A short time before Franklin Roosevelt met with Winston Churchill and Joseph Stalin at Tehran in 1943 he had a meeting with the Joint Chiefs of Staff aboard the battleship *Iowa*. The subject of the meeting was a British proposal to divide Germany after the war into three zones: a northwestern zone, to be governed by the British; a southwestern zone, to be run by the United States; and an eastern zone, which would include a jointly occupied Berlin, to be controlled by the Russians.

Roosevelt began the meeting by denouncing the British proposal. He told the chiefs that the United States should have the northwestern zone and not the southwestern zone. The northwest would be good for the ports at Bremen and Hamburg, and would give access to Norway and Denmark; the southwest would be good for nothing except trouble caused by instability in neighboring France, Belgium, and Luxembourg. In addition, Roosevelt told the chiefs, the United States, in the northwest zone, "should go as far as Berlin." "The United States," the President said firmly, "should have Berlin."

The Joint Chiefs of Staff replied that Roosevelt's plan was not feasible. He had neglected a simple fact, they said. Ever since the beginning of the war the United States had based its operations in Northern Ireland and southern and southwestern England. British forces had been based in northern England. Thus, logistics demanded that the British advance on Germany from the north, while the Americans advanced from the south.

But the President was not persuaded. He argued that the American army could easily move into the northwest once it entered Germany. In any case, the army would have to reach Berlin. There would be a race for Berlin among the Allies, who knew that control of the city would strengthen their hand in postwar negotiations, and the United States would have to get there first.

Before the generals could respond, Roosevelt pulled out a *National Geographic* map of Germany and began drawing lines on it. Vertical lines, horizontal lines, straight lines, and squiggly ones. Finally, he finished. There, he said, pointing at the map, is how Germany ought to be divided. The map showed the United States occupying a huge zone in the northwest including Berlin, with Britain getting a slightly smaller section in the southwest and the Soviet Union an even smaller area in the east.

The Joint Chiefs looked at each other in dismay, but said little. The President of the United States had formulated a policy and there was nothing they could do about it. The United States would have the northwest and Berlin.

Over the next four months the United States held negotiations with Britain and Russia on the question of the division of postwar Germany. But the negotiations were conducted entirely by the State Department—and the Joint Chiefs had neglected to tell Foggy Bottom what the President had decided about postwar occupation. So the State Department negotiated in ignorance. The British knew what Roosevelt's ideas were because the President had casually mentioned them at Tehran. But the State Department did not. The secretary of state might have thought of asking the President for his views, but the secretary was ill and his department was being run by an incompetent assistant secretary.

What about the *National Geographic* map that showed precisely what the President desired? After the meeting on the *Iowa* the President had handed it to General George Marshall. Marshall, in turn, had given it to Major General Thomas T. Handy, chief of the war department's Operations Division. And Handy had prudently put the map in the top-secret archives at the Pentagon—where it was promptly forgotten. As Handy recalled later, "To the best of my knowledge we never received instructions to send it to anyone at the Department of State."

Meanwhile, the British and the Russians came to agreement on the British plan. When the State Department told this to the President, he exploded. "What are the zones in the British and Russian drafts and what is the zone we are proposing?" he asked. "I must know this in order that it conform with what I decided on months ago."

The officers at the State Department were bewildered. What had the President decided on months ago? An army officer had mumbled something about the President's dislike of the British plan, but aside from that they had heard nothing.

Notes flashed back and forth between the White House and the State Department. The department learned about the *Iowa* meeting and was shown the *National Geographic* map. Eventually, everything was cleared up.

But by then it was too late. Momentum for the British plan had grown to the point where it could not be stopped. Months before, the United States probably could have prevailed and imposed its own plan on the Allies. But not now. Finally, on May 1, Roosevelt capitulated.

The United States had suffered a major diplomatic defeat because of a few silly blunders. Most important, the American bungling had led to the adoption of a plan giving Russia a zone that included Berlin. It is, of course, not clear that the Allies would have given Berlin to the United States if the State Department had demanded the city. But the possibility is one of the most tantalizing might-have-beens of modern history. Fewer blunders in 1943, and the world might have been spared the Berlin crises of 1948 and 1961.

SOURCES: Cornelius Ryan, *The Last Battle* (New York: Simon & Schuster, 1966), pp. 140–65; Herbert Feis, *From Trust to Terror* (New York: Norton, 1970), pp. 27–34.

★ WORLD WAR WEARINESS ★

Inscription found by an American reporter at Verdun, France, in 1945:

> Austin White—Chicago, Ill.—1918
> Austin White—Chicago, Ill.—1945
> This is the last time I want to write my name here.

SOURCE: *Webster's Guide to American History* (Springfield, Mass.: Merriam, 1971), p. 506.

★ WORLD WAR II MOST EXPENSIVE ★ WAR IN U.S. HISTORY

The total cost of World War II to the American people, including payments for veterans' benefits and interest on debts, amounted to about $560 billion. The cost of all other American wars:

Revolutionary War	$149 million
War of 1812	$124 million
Mexican War	$107 million
Civil War (Union)	$8 billion
Spanish-American War	$2½ billion
World War I	$66 billion
World War II	$560 billion
Korean War	$70 billion
Vietnam War	$121½ billion
Persian Gulf War	$80 billion*

SOURCE: *Dictionary of American History*, rev. ed. (New York: Scribner's, 1976), p. 228.

* Most of this total was paid by our allies.

FROM DOO-WOP TO DISCO

★ ≈≈≈

"The whole country is one vast insane asylum and they're letting the worst patients run the place."
—ROBERT WELCH, FOUNDER OF THE JOHN BIRCH SOCIETY

★ SCRAPBOOK OF THE TIMES ★

- During Watergate, John Connally was indicted in the milk scandal. When he ran for president in 1980, critics suggested he was tainted. Connally responded: "Well, what about it? I was tried and acquitted. I never drowned anybody. I was never kicked out of college for cheating."

- To see their favorite stars, 82 million people went to the movies every week in 1946. A decade later, because of television, only half as many people visited movie theaters weekly.

- As president, according to one White House servant's report, Harry Truman washed his own underwear.

- Amidst the crisis in the Middle East in 1948, U.S. ambassador to the United Nations Warren Austin made the helpful remark that he hoped Arabs and Jews would settle their differences "like good Christians."

- The last vice president born in a log cabin was Alben Barkley, who served under Harry Truman.

- Learning that "V.P." stands for "vice president," Alben Barkley's ten-year-old grandson coined a word when he suggested: "Why not stick in a couple of little e's and call it 'Veep'?"

- Confident of a Republican victory in 1948, the Republican Congress provided a large appropriation for the presidential inauguration, only to see it put to use by Harry Truman. The appropriation included $80,000 for grandstands alone.

- The year 1949 was the first year of the twentieth century in which a Negro was not lynched.

- In 1950, J. Edgar Hoover began the practice of issuing a list of the "Ten Most Wanted Criminals," after the idea was suggested to him by a friend of the fashion designer who invented the survey of the "Ten Best Dressed Women."
- Nebraska Senator Kenneth Wherry had a knack for mispronouncing and twisting words. Once he spent an hour on the Senate floor talking about the crisis in "Indigo-Chino," and another time he referred to the Joint Chiefs of Staff as the "Chief Joints of Staff." His malapropisms were known in the Senate as "wherryisms."
- When the mother of Dwight Eisenhower discovered her son reading books about war, she took them away and stowed them in the attic. Mrs. Eisenhower was a pacifist.
- Disgusted with bureaucratic intrigue and small-minded meddling, Albert Einstein remarked in 1954 that if he had his life to live over again he would want to be a plumber or a peddler—for the independence.
- Gary Cooper made this statement before the House Un-American Activities Committee when asked what he thought of communism: "From what I hear, I don't like it because it isn't on the level."
- After General Motors hired seven psychologists in the middle 1950s to determine the effect of Chevrolet's sounds on potential buyers, one Chevy general manager boasted: "We've got the finest door-slam this year we've ever had—a big-car sound."
- In 1957, Ford spent $20 million advertising a single car, the Edsel—$10 million before the car went on the market and another $10 million after it initially failed to sell.
- At political rallies young Congressman Lyndon Johnson would flamboyantly whip off his twenty-five-dollar Stetson hat and throw it to the crowd. He always got it back, however, because he would pay a small boy one dollar in advance of each rally to retrieve it.
- The New York *Times*'s endorsement of Kennedy for president in 1960 was surprising, since the paper usually supported Republicans. After the election Kennedy remarked: "In part,

at least, I am one person who can truthfully say, 'I got my job through the New York *Times*.'"

- When President Johnson found he had nothing interesting to do on a Sunday, he would call a small group of reporters to the White House and spend the afternoon complaining that he had no time as president to do half the great things he needed desperately to do.

- When Georgia Congressman Carl Vinson received an award from the American Political Science Association, he immediately ordered an investigation of the organization, since he had never heard of it.

- The last state to end prohibition was Mississippi, which adopted a local-option law in 1966.

- After charging in the 1968 campaign that Hubert Humphrey was soft on communism, Spiro Agnew retracted the comment when he was told it was the kind of remark made during the McCarthy era. Agnew explained that he had never heard anyone use the expression "soft on communism."

- Spiro Agnew's election as vice president of the United States in 1968 came just ten years after his first political victory—his election as vice president of a local Kiwanis club.

- The plaque which Neil Armstrong placed on the moon in 1969 was marred by an error. Instead of reading, "July A.D. 1969," it read, "July 1969 A.D."

- J. Edgar Hoover refused to allow people to walk on his shadow.

- Gerald Ford made an unsuccessful attempt to impeach Supreme Court Justice William O. Douglas in 1970.

- Nixon's 1972 inaugural parade was made pigeon-proof, through the use of a chemical sprayed on Pennsylvania Avenue.

- After the Russian wheat deal Lester G. Maddox of Georgia remarked: "The Communists got our grain, the Administration got credit for the deal, the speculators got the profit, and the rest of us got the bill."

- In 1973, President Nixon appointed William Saxbe attorney general of the United States—the same William Saxbe who on one occasion had referred to Haldeman and Ehrlichman as "those two Nazis" and another time had remarked, "I don't know whether or not [the Nixon administration's] the most corrupt but it's one of the most inept" in American history. After Nixon ordered the Christmas bombing of Haiphong Harbor, Saxbe had commented: "He's out of his fucking mind."

- Before Chuck Colson was "born again," the following Green Beret slogan hung over his bar at home: "If you've got 'em by the ——, their hearts and minds will follow."

- The White House Plumbers received their name when the grandmother of one of the men remarked, after reading in the New York *Times* that her grandson had been working on leaks, "Your grandfather would be proud of you, working on leaks at the White House. He was a plumber."

- In 1974 the Texas legislature repealed a law passed in 1837 that allowed a husband to murder his wife's paramour if he caught the pair in the act of lovemaking.

- Richard Nixon to Len Garment, his law partner: "You're never going to make it in politics, Len. You just don't know how to lie."

★ THE PECULIAR ORIGINS OF THE POINT ★ FOUR PROGRAM

One evening in 1948, after the day's work was done, Louis Halle, a low-level officer at the State Department, had a conversation with his superior, the deputy director of American Republic Affairs. Halle and the deputy spoke of nothing particularly serious, but Halle did mention an idea he had—an idea that would eventually become known as Truman's Point Four—American technical assistance for underdeveloped countries. The idea was not exactly original—the United States had for many years been giving technical aid to the countries of Latin America. But Halle had in mind a program of worldwide aid. The deputy director said the idea was worth considering.

In November the State Department's director of public affairs called a meeting to consider ideas for the President's upcoming inaugural address. At the meeting was Halle's superior, the deputy director. Quickly, everyone agreed that the President would have to cover three points. He would have to announce his full support for the embryonic United Nations; he would have to assure American allies that the European Recovery Program would be continued; and, finally, the President would have to declare that the United States favored the establishment of a defense organization for Western Europe.

Anything else? The director of public affairs looked searchingly around the room; the people present looked back blankly. Suddenly, the deputy director came to life. In a corner of his mind he recalled his conversation with Halle. What about having the United States give technical assistance to underdeveloped countries around the world? The public affairs director thought for a moment and then told the secretary to write down the idea. That would make four points, a good number to have. The meeting ended.

Over the next few weeks the "points" were sent to the appropriate offices at the State Department. Points one, two, and three passed with flying colors, but not point four. It was dropped entirely. Who knew what "technical assistance" meant? Which countries would aid be given to? At what cost? No one knew.

Points one, two, and three were sent to the White House.

Several days later one of the President's speechwriters telephoned the director of public affairs and complained about the three points. They were fine, he said, but they were not very exciting. Was there some other idea that might arouse a bit of attention?

The director paused for a few seconds. There was a fourth point, he said finally, but it had been dropped. What was it, the presidential assistant asked. The director told him. Fine, was the response, that will do just fine. The fourth point had been resurrected.

On January 20, President Truman delivered his inaugural address, with the four points. Instantly, newspaper reporters jumped on the last point. The other points were dull, but Point Four was a headline story. But what did it mean? Could the reporters see the plans, the blueprints? What did America's allies think about it? Had they suggested the idea?

The State Department, the White House, and the President

responded with stares. *They* certainly did not know what Point Four was. Nobody did. It was simply an eye-catcher, a professional speechwriter's device—not a plan.

The plain fact was that after the speechwriter's telephone conversation with the public affairs director, not one person had given another thought to the idea. For all intents and purposes the idea had not got beyond the teacup stage—the stage it had been in when Halle casually mentioned it to his superior.

Six days after his inaugural address Truman was asked about the origin of Point Four. The President responded: "Point Four has been in my mind, and in the minds of the government, for the past two or three years, ever since the Marshall Plan was inaugurated. It originated with the Greece and Turkey propositions. Been studying it ever since. I spend most of my time going over to that globe back there, trying to figure out ways to make peace in the world."

"Been studying it ever since," Truman said. But it took a full twenty-two months for the plan to be put into operation.

SOURCE: Louis J. Halle, *The Society of Man* (New York: Dell, 1969), chap. 1.

★ TRUMAN AND THE GREAT SEAL ★

One day Harry Truman walked into the Green Room at the White House and looked down at the rug. Something bothered him about that rug. Finally, he realized that the head of the eagle on the presidential seal in the center of the rug was turned the wrong way. Instead of facing the olive branch, this eagle's head pointed toward the arrows. Promptly Truman ordered that the rug be restitched with the eagle's head turned the right way.

SOURCE: Lilian Rogers Parks, *My Thirty Years Backstairs at the White House* (New York: Fleet, 1961), p. 345.

★ LBJ GOES TO CONGRESS ★

"Landslide Lyndon" earned his nickname and his United States Senate seat in 1948 after winning the Texas Democratic primary by an overwhelming eighty-seven votes. There were almost a million votes cast in the election. The future president's margin of victory, however, was closer than mere statistics indicate.

In Jim Wells County, Johnson won by over 1,000 votes. After a statewide recount, however, election officials in the south Texas county reported that Johnson had actually received over 1,200 votes. In the recount 202 votes that officials had previously miscounted suddenly appeared, all in the same handwriting, with the same ink, and in alphabetical order. All 202 names were on Johnson's election sheet. His opponent, former Governor Coke R. Stevenson, charged fraud, but a brief investigation upheld the election, and "Landslide Lyndon" was declared the winner, 494,191 to 494,104.

One former Johnson aide later commented on the 1948 Senate race. "Of course," he revealed, "they stole that election. That's the way they did it down there. In 1941, when Lyndon ran the first time for the Senate, he went to bed one night thinking he was 5,000 votes ahead . . . and he woke up next morning 10,000 votes behind. He learned a thing or two between 1941 and 1948."

Reportedly, one name that was recorded as a vote for Johnson in 1948 belonged to the grandfather of William F. Buckley Jr., the New York conservative. Even though the elder Buckley had died in 1904, the old Texan apparently considered LBJ's election important enough to rise from the grave to cast his ballot. Said Buckley Jr., of his grandfather's voting record: "I am very proud of my grandfather's sense of civic obligation. Clearly it runs in the family."

SOURCES: *Newsweek*, August 8, 1977, p. 27; *National Review*, September 2, 1977.

★ THE SEARCH FOR THE LAST REBEL YELL ★

In the spring of 1949, Frank Tolbert, famous Texas historian and folklorist, went searching with a tape recorder for what legend describes as the most fearsome sound ever uttered by man—the Rebel yell of the soldiers of the Confederate army. His hunting ground was Texas, which at that time had four living Civil War veterans, all Confederates. Time was of the utmost importance, since all four men were over a hundred years old.

Tolbert first visited Joseph Haden "Uncle Hade" Whitsett, a 103-year-old retired gentleman farmer.

"Can't do it," Uncle Hade informed him. "Can't Rebel yell. I'm sorry. I tried to learn it a thousand times when I was with [General Joseph] Shelby's Escort during the war. I didn't seem to have the right kind of voice. Wish you could have heard Uncle Joe, himself, Rebel yell. He could make a full-voiced loafer wolf sound like someone blowing on a penny whistle." Then, indicating Tolbert's tape recorder, Uncle Hade continued, "I can't do you no Rebel yell, but I'll sing into your gadget."

Walt Williams, 107 years of age, was the person Tolbert next interviewed.

"Used to could do it," he replied. "But I haven't got the throat linings for it now. When you get a hundred seven you can't do everything you want no more."

Williams then offered Tolbert his secret for a long life. "Don't et much. When I was riding up the Chisholm Trail the range cooks sort of held it against me because I was such a light-eating man."

Tolbert next traveled to Wichita Falls, the home of Thomas E. Riddle, a 104-year-old veteran who had recently divorced his third wife and claimed to be looking for a fourth.

"Can you do the Rebel yell?" inquired Tolbert.

"Yes, I'm feeling pretty well," Riddle answered. "I'm peart enough to walk to the courthouse and back every day."

Tolbert shouted his question.

"I'll be a suck-egg mule!" Riddle exclaimed. "I'm sure getting deaf before my time. No, sir, I can't do the Rebel yell—not anymore. Not rightly anyway. I remember hearing it the best at Gettysburg and in the Wilderness. It was a terrible sound. But I can't do it rightly anymore. No old man can. It's a young man's yell. For those seventy or under."

Samuel Merrill Raney, 103, was Frank Tolbert's last hope. In introducing himself, Tolbert mentioned that he was from Dallas.

"Haven't been to Dallas in sixty year," replied Raney. "I went down there the last time to help pave the streets with *bois d'arc* blocks. I suppose it has growed. Last time I was in Dallas it was full of carpetbaggers. Here it still is."

A tall, athletic-looking, white-haired man came in from the field, leaped over a wire fence, and trotted up to the porch. He was George W. Raney, the seventy-five-year-old "baby" of the family, Samuel's son.

"Can you do the Rebel yell?" Tolbert asked Samuel.

George Raney answered, "Papa can. But he best not."

Samuel waved his son silent. He spoke of the war and began to reminisce about his first action at the Battle of Murfreesboro. The cedar trees and cotton stalks set afire by the roaring cannon . . . the band playing . . . the charge. Abruptly the old veteran threw back his head and started yelling, "like an opera singer hitting almost impossibly high notes, . . . as if a mountain lion and a coyote were crying in chorus." As the yell trailed off into convulsive coughs, George Raney ran to the well to fetch his father a dipper of water, while Tolbert headed for his car and the tape recorder.

"With this thing we can make a record of that yell," Tolbert explained. "I'll give you a record of it, and you can play it for people who want to hear a real Rebel yell. All I've got to do is plug in this switch and then you can start yelling again."

"We haven't got no electricity," said George Raney. "We never did tie on to REA [Rural Electrification Administration]."

"We could get into the car and drive to the nearest house that has electricity," suggested Tolbert.

"Best not," answered George firmly. "Papa don't ride in cars."

"Never do," agreed the old man. "Cars shake me up too much. Bother my kidneys."

"You come back later and bring a battery for that talking machine and he'll yell for you," George recommended.

A few days later Tolbert returned to Raney's farm armed with a battery-powered tape recorder. A big red hen was resting on Raney's rocking chair in the breezeway. Tolbert knocked on the bedroom door, but there was no answer.

"Mr. Raney, Mr. Raney!" he shouted.

George Raney suddenly appeared from the field, hurdled the wire fence, recognized Tolbert, and remorsefully informed him: "Papa died."

Source: Frank X. Tolbert, *An Informal History of Texas* (New York: Harper & Brothers, 1961), pp. 237–44. Dialogue reprinted by permission of Harper & Row, Publishers, Inc.

★ There Are No Smiling Russians ★

In the days of McCarthyism, spokesmen against communism could say almost anything about the Russians and be believed. But there were limits to people's credulity even then, as the famous libertarian Ayn Rand discovered.

In October 1947 she testified before the notorious House Un-American Activities Committee about the 1944 film *Song of Russia*. She criticized the movie for showing the Russian people smiling and remarked that she had "never seen so much smiling in my life, except on the murals of the World's Fair pavilion of the Soviet. If any one of you have seen it, you can appreciate it. It is one of the stock propaganda tricks of the Communists, to show these people smiling."

This comment sparked the following remarkable exchange:

Congressman John McDowell: You paint a very dismal picture of Russia. You made a great point about the number of children who were unhappy. Doesn't anybody smile in Russia any more?

Miss Rand: Well, if you ask me literally, pretty much no.

Mr. McDowell: They don't smile?

Miss Rand: Not quite that way; no. If they do, it is privately and accidentally. Certainly, it is not social. They don't smile in approval of their system.

Source: U.S. Congress, House, Committee on Un-American Activities, *Hearings on Hollywood* (80th Cong., 1st sess., October 20–24, 27–30, 1947 [Washington: Government Printing Office, 1947]).

★ A Moment in the Life of HUAC ★

Paul Robeson, the accomplished Negro singer, was an unabashed apologist for Soviet Russia. In the summer of 1956 he was called to testify before the House Un-American Activities Committee.

Staff Director Richard Arens: Are you now a member of the Communist Party?

Mr. Robeson: Would you like to come to the ballot box when I vote and take out the ballot and see?

Mr. Arens: Mr. Chairman, I respectfully suggest that the witness be ordered and directed to answer that question.

The Chairman: You are directed to answer the question.

Mr. Robeson: I stand upon the Fifth Amendment of the American Constitution.

. . .

Mr. Arens: Now, tell this Committee whether or not you know Nathan Gregory Silvermaster.

Mr. Robeson: [Laughter.]

Congressman Gordan Scherer: Mr. Chairman, this is not a laughing matter.

Mr. Robeson: It is a laughing matter to me, this is really complete nonsense.

Mr. Arens: Have you ever known Nathan Gregory Silvermaster?

Mr. Robeson: I invoke the Fifth Amendment.

Mr. Arens: Do you honestly apprehend that if you told whether you know Nathan Gregory Silvermaster you would be supplying information that could be used against you in a criminal proceeding?

Mr. Robeson: I have not the slightest idea what you are talking about. I invoke the Fifth—

Mr. Arens: I suggest, Mr. Chairman, that the witness be directed to answer that question.

The Chairman: You are directed to answer the question.

Mr. Robeson: I invoke the Fifth.

Mr. Scherer: The witness talks very loud when he makes a speech, but when he invokes the Fifth Amendment I cannot hear him.

Mr. Robeson: I invoked the Fifth Amendment very loudly. You know I am an actor, and I have medals for diction.

Mr. Scherer: Will you talk a little louder?

Mr. Robeson: I can talk plenty loud, yes. [Almost shouting now.] I am noted for my diction in the theater.

Mr. Ahrens: Who are Mr. and Mrs. Vladimir P. Mikheev? Do you know them?

Mr. Robeson: I have not the slightest idea, but I invoke the Fifth Amendment.

. . .

Mr. Robeson: To whom am I talking?

The Chairman: You are speaking to the Chairman of this Committee.

Mr. Robeson: Mr. Walter?

The Chairman: Yes.

Mr. Robeson: The Pennsylvania Walter?

The Chairman: That is right.

Mr. Robeson: Representative of the steelworkers?

The Chairman: That is right.

Mr. Robeson: Of the coal-mining workers and not United States Steel, by any chance? A great patriot.

The Chairman: That is right.

Mr. Robeson: You are the author of all of the bills that are going to keep all kinds of decent people out of the country.

The Chairman: No, only your kind.

Mr. Robeson: Colored people like myself, from the West Indies and all kinds. And just the Teutonic Anglo-Saxon stock that you would let come in.

The Chairman: We are trying to make it easier to get rid of your kind, too.

Mr. Robeson: You do not want any colored people to come in?

The Chairman: Proceed.

SOURCE: U.S. Congress, House, Committee on Un-American Activities, *Hearings on June 12, 1956* (84th Cong., 2d sess. [Washington: Government Printing Office, 1956]).

★ Gold Counted in Fort Knox ★

One of the least serious charges ever made against the New Deal Democrats was that they had stolen the gold in Fort Knox. Only a few crackpots believed the accusation. But in 1953 the Daughters of the American Revolution forced Dwight Eisenhower, the first Republican president in twenty years, to have the gold counted. Investigators found that the Fort contained $30,442,415,581.70 worth of the precious metal. That was ten dollars less than it should have been. Mrs. Georgia Clark, treasurer of the United States under the Democrats, sent the government a check to cover the loss.

SOURCE: Eric F. Goldman, *The Crucial Decade—And After: America, 1945–1960* (New York: Vintage, 1960), p. 239.

★ Howard Hunt and the Communists ★

In 1954, E. Howard Hunt was in Guatemala helping overthrow the anti-American government that had recently come to power. The future Watergate burglar was one of the leading CIA agents involved in the coup against the leftist government. Just before he departed from the country, Hunt revealed a soft streak and ordered a small band of prisoners set free rather than shot.

Afterwards Hunt told friends that that was the greatest mistake of his life. One of the prisoners he freed, he later learned, was Ché Guevara, the Cuban revolutionary. Hunt swore that he would never again succumb to feelings of compassion.

SOURCE: Douglas Hallett, "A Low-Level Memoir of the Nixon White House," *New York Times Magazine*, October 20, 1974, p. 39.

★ THE DAY AMERICA DIDN'T PULL THE ★ TRIGGER ON VIETNAM

On the morning of April 3, 1954, a Saturday, Secretary of State John Foster Dulles summoned eight leaders of Congress to the State Department for a secret meeting. The secretary began by striking an authoritative note. "The President," he said, "has asked me to call this meeting." The legislators fidgeted nervously. Dulles went right to the point. The situation in French Indochina, Dulles declared, had become critical. Thus, the President wanted a joint congressional resolution allowing him to use air and naval forces to save the French.

Arthur Radford, chairman of the Joint Chiefs of Staff, stepped forward. He pointed to a map on the wall. Dienbienphu had been under siege for three weeks. At any moment it might fall. As a matter of fact, it may already have fallen—communications were so bad the administration was unsure what was happening. And if Dienbienphu fell, Indochina would fall, and with it Free Asia.

Dulles again. If America did not act and act fast, the West would be forced out of Asia. The French barely had enough strength to evacuate their men. They could not be relied on.

Radford once more. Two hundred planes from the carriers *Essex* and *Boxer*, and more planes from the Philippines, could do the job. A neat, single strike and Dienbienphu could be rescued.

Then the questions began. Would this mean war? Yes. What if the neat, single strike did not work? Well, the United States would have to find some other way to save Dienbienphu. Would land forces be resorted to? Radford wouldn't say. Then a very important question: "Does this plan have the approval of the other members of the Joint Chiefs of Staff?"

"No," said Radford.

"How many of the three agree with you?"

"None."

"How do you account for that?"

"I have spent more time in the Far East than any of them and I understand the situation better."

Lyndon Johnson, the minority leader of the Senate, asked a few questions. Had Dulles sought the support of other nations? Had he gotten it? Would other countries help in the fighting? Johnson remarked that the United States certainly would want to avoid another Korea, that it would be absolutely wrong to try to go it alone. Dulles answered that he had not asked any other country for help.

Suddenly the room exploded with questions. No allies? What about the United Nations? And Britain? Finally, the meeting came to a close, all the legislators agreeing on one central point. If the United States could not get allies, it should not go to war.

Dulles spent two weeks looking for allies but could not find any. Most important, he could not persuade Britain to support the proposed United States action.

Lyndon Johnson's questions had undermined Dulles's entire plan—no allies, no war. In the middle of March the members of the National Security Council had voiced the same concern as LBJ. But Dulles had ignored the council's opinion, as he safely could. But Dulles could not ignore Johnson. Without him, there could be no war.

A decade later, however, war came, under the leadership of the man who, more than anyone else, was responsible for keeping the United States out of war in 1954, Lyndon Johnson.

SOURCE: Chalmers Roberts, "The Day We Didn't Go to War," *Reporter*, September 14, 1954, pp. 31–35.

★ VERTICAL INTEGRATION ★

Integration posed great problems for some Southern communities. In Danville, Virginia, blacks had never used the main public library. Until the 1954 Brown decision, no one in Danville had ever considered the possibility that they might want to use the all-white building. Now the Supreme Court had ruled race discrimination in public facilities unconstitutional. The main library's "whites only" policy must be illegal, but no one knew exactly what real-life changes the high-court ruling might precipitate. Blacks had never tried to enter the building. Danville librarians, as well as the rest of the nation, would have to wait and see how the integration decision would affect America and the South.

But not every American waited. Harry Golden, founder and editor of the Charlotte, North Carolina, *Carolina Israelite*, proved more than willing to publish ideas designed to ease the court-ordered mixture of races. One of his more brilliant visions for school integration was described in an article entitled "The Vertical Negro Plan." The idea was simple. Whites and blacks throughout the South stood at the same grocery and supermarket counters; they deposited money at the same bank teller's windows; they walked through the same dime and department stores; they paid phone and light bills to the same clerk; they even stood at the same drugstore counter together. Only when a black stopped being "vertical" and tried to sit down did Southern whites start complaining. A black who served tables in an elegant restaurant was more than welcome, as long as he remained on his feet and did not attempt to receive service himself. Obviously, then, wrote Golden, the state legislature should pass an amendment providing only desks for public schools. No chairs. With only "vertical Negroes" integrating a school, no Southerner could possibly have any basis for complaint. In addition, millions of dollars would be saved on the cost of chairs.

A short while after the publication of Golden's article, conflict began at the Danville Public Library. One afternoon a small group of high school–age blacks entered the library and exercised their civil rights by actually studying at one of the large desks in the center of the building. At closing time they quietly put down their work and left. They did not even try to check out any books.

The building was closed for several months while library officials considered possible solutions to their problem. No one wanted to alienate any longtime users. On the other hand, the

The Danville Public Library.

high-court ruling was very clear concerning a matter such as this. Finally, in a desperate attempt to please everyone, the library reopened—minus every chair in the building.

A week or two later someone must have informed library officials that the "Vertical Negro Plan" had been a spoof. The chairs were quietly and unceremoniously returned. Danville blacks, however, remained welcome not only to use the facility, but even to take out books.

SOURCE: Lecture by Payton McCrary at Vanderbilt University, March 1978.

★ AMERICA CHOOSES AN AMBASSADOR ★

Ambassadorships are always assigned to the best possible candidates. Certainly President Dwight Eisenhower had only this in mind when he nominated chain-store president Maxwell H. Gluck to be America's ambassador to Ceylon. Gluck, not famous for his experience in international politics or diplomacy, was better known for personally contributing some twenty to thirty thousand dollars to the Republican 1956 campaign chest. Senate questioning of Gluck during his confirmation hearings was led by William Fulbright.

Mr. Fulbright: What are the problems in Ceylon you think you can deal with?

Mr. Gluck: One of the problems are the people there. I believe I can—I think I can establish, unless we—again, unless I run into something that I have not run into before—a good relationship and good feeling toward the United States.

Mr. Fulbright: Do you know our Ambassador to India?

Mr. Gluck: I know John Sherman Cooper, the previous Ambassador.

Mr. Fulbright: Do you know who the Prime Minister of India is?

Mr. Gluck: Yes, but I can't pronounce his name. [The name was Nehru.]

Mr. Fulbright: Do you know who the Prime Minister of Ceylon is?

Mr. Gluck: His name is unfamiliar now, I cannot call it off.

Reporters later asked Eisenhower if Gluck's strongest credential for the ambassadorship had not in fact been his thirty-thousand-dollar donation. "Unthinkable," snapped Ike. "Now, as to the man's ignorance. . . . Of course, we knew he had never been to Ceylon, he wasn't thoroughly familiar with it; but certainly he can learn."

After one year in Ceylon, Gluck resigned.

SOURCE: Richard Hofstadter, *Anti-Intellectualism in American Life* (New York: Vintage, 1963), p. 10.

★ Ike at Gettysburg ★

Intellectuals and President Eisenhower were not cut from the same mold. Ike's definition of an intellectual was "a man who takes more words than is necessary to say more than he knows." Intellectuals countered by equating Eisenhower with the mundane and complacent "Leave It to Beaver" 1950s America, a place where, as satirist Jules Feiffer remarked, "satire doesn't stand a chance against reality any more." In 1957, Doris Fleeson, a Washington reporter, rewrote the Gettysburg Address as Ike would have delivered it had he been there after the battle instead of Lincoln:

"I haven't checked these figures, but 87 years ago, I think it was, a number of individuals organized a governmental setup here in this country, I believe it covered certain eastern areas, with this idea they were following up, based on a sort of national independence arrangement. . . .

"[W]e can't sanctify this area—we can't hallow, according to whatever individual creeds or faiths or sort of religious outlooks are involved, like I said about this particular area. . . . The way I see it, the rest of the world will not remember any statements issued here, but it will never forget how these men put their shoulders to the wheel and carried this idea down the fairway."

SOURCE: Eric F. Goldman, *The Crucial Decade—And After* (New York: Vintage, 1960), pp. 304–5.

★ Presidents Are Really Not Needed ★

Republican Senator Kenneth Keating: "Roosevelt proved a man could be President for life; Truman proved anybody could be President; and Eisenhower proved you don't need to have a President."

SOURCE: Leon Harris, *The Fine Art of Political Wit* (New York: Dutton, 1964), p. 228.

★ SPEAKER RAYBURN'S TACT ★

In the 1950s the reigning monarch of the House of Representatives was Texan Sam Rayburn. When Daniel K. Inouye, a Hawaiian of Japanese descent who lost an arm during World War II, came to Washington in 1959 as his state's first congressman, he sought out Speaker Rayburn. Inouye introduced himself formally to the Speaker as Daniel K. Inouye of Hawaii. Rayburn replied, "I know who you are. How many one-armed Japs do you think we have around here?"

SOURCE: Steven Gerstel, newspaper reporter, Washington, D.C., in a personal interview with Richard Shenkman at the Capitol, January 1979.

★ MODERN AMERICAN JUSTICE ★

Critics often claim that the modern American penal system is a tortuous bureaucratic maze where justice is rarely realized. In 1925 sixteen-year-old Stephen Dennison of upper New York State was convicted of stealing a five-dollar box of candy. Sentenced to the state reformatory, he was later moved to the state penitentiary. There he broke minor rules, which added years to his term. In 1959 he finally gained his release, *thirty-four* years after committing only one crime, stealing a box of candy.

Seven years later the Court of Appeals of New York awarded Dennison $115,000 in an attempt to compensate him for his mistreatment. The court stated, however, that "no amount of money could compensate Dennison for the injuries he suffered and the scars he bears."

SOURCE: Jay Robert Nash, *Bloodletters and Badmen* (New York: M. Evans and Company, 1973), p. 153.

★ THE SANE GOVERNOR OF LOUISIANA ★

Few modern governors must be declared sane to keep their job. But in 1959 Governor Earl Long of Louisiana, brother of the late demigod Huey Long, was forcibly kidnapped from office and sent to a mental institution. Less than four weeks later, through sheer determination and skillful political maneuvers, he was back in office—sanely firing those responsible for his incarceration.

Long's health had seriously declined during the May legislative session. A bill to disfranchise blacks and poor whites had tremendous support, and many have suggested that Long's untiring efforts to block removals from voter registration lists led directly to his health problems. Others claim that the disfranchisement battle merely coincided with the governor's failing health.

On May 27, Long delivered two long-winded diatribes against his political opponents. His physical health, he later freely admitted, was not good. Earl's wife, Blanche, however, thought her husband was on the verge of lunacy. She summoned doctors and relatives to the Governor's Mansion, and together they decided Earl needed immediate psychiatric attention and should be sent to a mental clinic in Galveston, Texas. When the medics arrived to take Long away, the governor battled furiously. Even after he was strapped to a stretcher and given sedatives, six men were required to keep Long in place. Blanche signed a formal petition requesting the governor's treatment to prevent Earl's leaving the hospital.

Governor Long fought the proceedings to have him declared insane, signing his name and describing himself "Governor in exile, by force and kidnapping." From his hospital window he shouted to reporters, "I'm no more crazy than you are." But Long also worked at getting out of the Galveston hospital. He became a model patient and, within about two weeks, Blanche withdrew any objection to her husband's release. Earl voluntarily agreed to enter a clinic in New Orleans.

Less than twenty-four hours after his arrival in New Orleans, Governor Long walked out of the clinic, got in a car, and started driving toward Baton Rouge, the state capital. His wife, Blanche, however, raced ahead of him and made arrangements to have the governor committed to the Southeast Louisiana Hospital at Mandeville. When Long reached the capital, he was "examined" by the East Baton Rouge Parish coroner and a local psychiatrist for forty-five minutes, then committed by a district judge who had been one of Long's political rivals. The governor was taken by police to Mandeville.

For eight days Long plotted his release. A habeas corpus hearing to decide on his release was scheduled for June 26. On the eve of the hearing Earl filed a separation suit against his wife. Legally, she could no longer obtain her husband's commitment to any hospital. A tape recording of Long, in which he personally declared he had not been up to par physically but had remained mentally sound, was released to a New Orleans radio station. With the aid of the lieutenant governor and a loyal state senator, Long's attorney called a meeting of the state hospital board, the board that in a few days would hear the governor's case, and fired all members who Long felt might hesitate to declare him sane.

The hearing on June 26 lasted fifteen minutes, with the newly appointed superintendent of hospitals, Jess H. McClendon, saying of Earl, "He should be released and I intend to do so." McClendon, before his appointment, had been a lifelong friend of the governor. The state attorney general joined in the release motion, and Earl, found sane, returned to his duties as Louisiana's governor. His marriage to Blanche, however, was permanently disrupted. They remained separated until Long's death in 1960.

SOURCE: Richard B. McCaughan, *Socks on a Rooster: Louisiana's Earl K. Long* (Baton Rouge: Claitor's Bookstore, 1967), pp. 170–78.

★ JFK POKES FUN AT LBJ ★

At a Gridiron Dinner in 1958, Senator Kennedy told the correspondents about a dream he had had concerning the presidency: "I dreamed about 1960 myself the other night and I told Stuart Symington and Lyndon Johnson about it in the cloakroom yesterday. I told them how the Lord came into my bedroom, anointed my head, and said, 'John Kennedy, I hereby appoint you President of the United States.' Stuart Symington said, 'That's strange, Jack, because I too had a similar dream last night in which the Lord anointed me and declared me, Stuart Symington, President of the

United States and Outer Space.' Lyndon Johnson said, 'That's very interesting, gentlemen, because I too had a similar dream last night and I don't remember anointing either of you.'"

SOURCE: Leon A. Harris, *The Fine Art of Political Wit* (New York: Dutton, 1964), p. 261.

★ KENNEDY ON NIXON ★

- When Nixon reputedly accused Kennedy of telling "a barefaced lie," JFK commented: "Having seen [him] four times close up . . . and made up, I would not accuse Mr. Nixon of being barefaced, but the American people can determine who is telling the truth."
- "Do you realize the responsibility I carry?" Kennedy asked his supporters during the campaign. "I'm the only person between Nixon and the White House."
- "Last Thursday night," Kennedy said during the campaign, "Mr. Nixon dismissed me as 'another Truman.' I regard that as a great compliment. I consider him another Dewey."

SOURCE: Theodore Sorensen, *Kennedy* (New York: Harper & Row, 1965), pp. 208, 180, 185.

★ JIM CROW, REALTOR ★

In 1961, President Kennedy nominated Robert C. Weaver, later the first Negro cabinet member, to be head of the Housing and Home Finance Agency. At his confirmation hearings before the Senate Banking Committee in February, Weaver received nearly unanimous praise, but was pressed hard by William Arvis Blakley. Blakley was a new face on the committee, having just been appointed to the Senate to fill the vacancy caused by the election of Lyndon Johnson as vice president. Before he began his questioning, Blakley humbly told the committee that he was reluctant "to take a stand or bring out information," since he had had so little experience. But the chairman reassured him.

Blakley began his questioning by asking Weaver about certain black groups that had been named by the House Un-American Activities Committee as communist fronts. Had Weaver been a member of the National Negro Congress? Did he know it was a subversive organization?

No, replied Weaver, he had never been a member of the congress and, in any case, it did not become communist-tainted until 1939 or 1940. The congressman had asked about Weaver's connection with the organization in 1937.

Blakley then asked Weaver about the Washington Cooperative Book Shop. Had Weaver ever been a member of the shop? Weaver conceded that he had been, in order to get a 20 percent discount. But he had resigned shortly after learning that the shop was not run democratically. Had he ever gone to a membership meeting?

"Oh, yes; I was there," Weaver replied.

Blakley, satisfied that he had uncovered a skeleton in Weaver's closet, continued. Did Weaver know that one of his books, *The Negro Ghetto*, which explained how city and suburban realtors helped create ghettos, had been praised by the Worker's Book Shop of New York, a communist bookshop? Would he be pleased if the book were prominently displayed in a communist bookshop?

"Yes, and my publisher would appreciate it too."

Finally, Blakley asked Weaver about a review of *The Negro Ghetto* that had appeared in the August 1948 issue of *Masses and Mainstream*, the well-known communist magazine. "Did you know about this review?"

"I do not know whether I did or not, sir," said Weaver. "I read about 250 reviews. If it came to my attention, I probably did read it. Do you know who wrote it? I might be able to identify it better."

"Yes. This seems to be by J. Crow, realtor." (The review was bylined, "J. Crow, Realtor," but was signed at the bottom by Herbert Aptheker.)

"Who?"

"J. Crow, realtor. Do you know J. Crow?"

"I did not know he wrote book reviews."

"Yes, sir," said Blakley. "This book reviewer seems to have been J. Crow, realtor. He went under another name sometimes, I suppose."

That June, William Arvis Blakley was defeated in a special election. His six months in the U.S. Senate were over.

SOURCE: U.S. Congress, Senate, Committee on Banking and Currency, *Hearings on the Nomination of Robert Weaver* (87th Cong., 1st sess., February 7, 8, 1961 [Washington: Government Printing Office, 1961]).

★ EXTRA WORK FOR WILLIAM O. DOUGLAS ★

Supreme Court Justice William O. Douglas was a punctual, highly organized man. He was also the only man in the history of the Supreme Court to write both the majority and minority opinions for the same case.

Meyer et al. vs. *United States* (364 U.S. 410), a life insurance–payment case, was argued before the Court on October 12, 1960. The opinion of the nine justices was split, and Justice Charles E. Whittaker, an Eisenhower appointee who, by his own admission, lacked the decisiveness and stamina required by the large Court workload, was directed to write the majority opinion. Douglas was chosen to write the minority opinion.

Typically, Justice Douglas finished his work in a few days. Days soon turned into weeks, however, and Justice Whittaker's majority opinion was still not available. Douglas, of course, could not submit his minority report until the majority decision had also been circulated.

Over a month later Douglas visited Whittaker at his office to discuss a totally different matter and found his fellow justice worriedly pacing the floor. Whittaker simply could not, he confessed to Douglas, find the proper inspiration to begin writing the majority opinion in the *Meyer* case.

That's because you found for the wrong side! ribbed Douglas.

No, answered Whittaker, he had made the correct decision, but somehow the writing would not flow. Douglas then realized that he had an opportunity to speed up the legal process. Even though Douglas had sided with a minority of his fellow justices in the case, he understood completely the rationale of the majority's decision.

If you think it might help, he offered, I'll write a few notes to get you started.

Whittaker accepted this sympathetic proposal, and within a few hours he had Douglas's majority opinion on his desk. With scarcely any changes, Douglas's writing became the recorded opinion of the Court. Legal records, of course, credit Douglas with writing the minority opinion only.

SOURCE: Dagmar S. Hamilton, Austin, Texas, in a personal interview with Kurt E. Reiger, February 1979.

★ CATCH-18 ★

During the eight or nine years that Joseph Heller was writing his first novel, the working title of the book was "Catch-18." Then, just as the book was going into production, *Publishers Weekly*, a trade magazine, informed Heller that Leon Uris was coming out with a book with the same number in the title. The magazine warned that the public would probably not accept two books with similar titles, and opined that if people had to choose, they would pick the Uris novel, since Uris was a familiar name.

Heller immediately sank into depression. He even took a leave of absence from his job as a writer of advertisements for *McCall's* to brood about his problem. He had specially picked the number eighteen because it was the only multi-syllable number that begins with a vowel, except for the number eleven, which could not be used since it was part of the title of a recently released movie.

For four weeks Heller worried about the problem, until one day his editor called with an idea for a new title. The editor suggested "Catch-22," and instantly Heller agreed.

SOURCE: Joseph Heller, in a television interview with Bill Boggs on station WNEW, New York, March 15, 1979.

★ LBJ, Texan ★

Lyndon Johnson was the earthiest president in recent American history. In the 1950s, when Richard Nixon was vice president and LBJ was the majority leader of the Senate, reporters sensed that the Texan had not taken a particular speech of Nixon's very seriously. When asked why, LBJ smiled and said, "Boys, I may not know much, but I know the difference between chicken shit and chicken salad."

On another occasion LBJ thought of dismissing FBI director J. Edgar Hoover, then realized it would be too difficult. "Well," commented a chagrined Johnson, "it's probably better to have him inside the tent pissing out than outside pissing in."

SOURCE: David Halberstam, *The Best and the Brightest* (New York: Random House, 1972), p. 436.

★ J. Edgar Hoover Asks Who Sartre Is ★

In June 1964, FBI director J. Edgar Hoover started an investigation of Jean-Paul Sartre when he read a newspaper report that the famous French philosopher had joined the "Who Killed Kennedy Committee."

The newspaper identified Sartre as an author only. Hoover immediately issued an order: "Find out who Sartre is."

SOURCE: UPI, *Tennessean*, January 1978.

★ LBJ Learns to Reason Together ★

"Come now, let us reason together," became a catch phrase of the Lyndon Johnson administration. Today everyone associates the Isaiah 1:18 quote with the late president. But how did he learn to use it?

In LBJ's early days as a politician, he once had a strong disagreement with the head of a power company. The future president wanted a small Rural Electrification Administration line constructed in his county district in Texas. The power company said no. Finally, in total exasperation, LBJ told the power company official to go to hell.

After the verbal exchange, a friendly ex-senator offered Johnson some timely advice. "Telling a man to go to hell and then making him go is two different propositions. First of all, it is hot down there and the average fellow doesn't want to go, and when you tell him he has to go, he just bristles up and he is a lot less likely to go than if you hadn't told him anything. What you better do is get out the Good Book that your mama used to read to you and go back to the prophet Isaiah and read what he said. He said, 'Come now, let us reason together.'" From that time forward LBJ was quoting Isaiah and reasoning together.

Of course, what the old ex-senator and LBJ left out of the Bible quotation was Isaiah 1:19. Many Bible-reading contemporaries, however, quickly pointed to it as the quote most exemplified by the Johnson administration:

"If ye be willing and obedient, ye shall eat the good of the land;

"But if ye refuse and rebel, ye shall be devoured by the sword, for the Lord hath spoken."

SOURCE: Paul F. Boller Jr., *Quotemanship* (Dallas: Southern Methodist University Press, 1967), p. 917.

★ LBJ's Bathroom Talks ★

To Lyndon Johnson there was nothing strange about talking to someone while sitting on the Imperial Flusher. If anything Johnson thought it was odd when people stopped talking simply because a visit to the bathroom was necessary. He explained his ideas about the propriety of bathroom conversations to biographer Doris Kearns by way of an anecdote:

"[Once a Kennedyite came into the bathroom] with me and then found it utterly impossible to look at me while I sat there on the toilet. You'd think he had never seen those parts of the body before. For there he was, standing as far away from me as he possibly could, keeping his back toward me the whole time, trying to carry on a conversation. I could barely hear a word he said. I kept straining my ears and then finally I asked him to come a little closer to me. Then began the most ludicrous scene I had ever witnessed. Instead of simply turning around and walking over to me, he kept his face away from me and walked backward, one rickety step at a time. For a moment there I thought he was going to run right into me. It certainly made me wonder how that man had made it so far in the world."

SOURCE: Doris Kearns, *Lyndon Johnson and the American Dream* (New York: Harper & Row, 1976), pp. 241–42.

★ LBJ GOES PUBLIC ★

As a politician LBJ did a lot of traveling. Once he visited Thailand for a conference. At the conference Johnson was feasted royally and given plenty to drink, which made him a frequent visitor to the rest room. As he emerged from the rest room on one of these trips, he met a group of reporters. Instantly, Johnson opened his fly, pulled out his membranous Texan, and commented, "Don't see 'em this big out here, do they?"

SOURCE: Steven Gerstel, newspaper reporter, Washington, D.C., in a personal interview with Richard Shenkman at the Capitol, January 1979.

★ LBJ GIVES AWAY TOOTHBRUSHES ★

LBJ loved to give gifts. In his first year as President he spent over three times as much as his predecessor on gifts. He particularly liked giving electric toothbrushes. "I give these toothbrushes to friends," he explained to biographer Doris Kearns, "for then I know that from now until the end of their days they will think of me the first thing in the morning and the last at night."

To make doubly sure friends would not forget him, LBJ gave particular individuals more than one toothbrush. Doris Kearns received her first toothbrush when she worked as an intern at the White House in 1968. Over the next ten years she received more than twelve toothbrushes from Johnson.

SOURCE: Doris Kearns, *Lyndon Johnson and the American Dream* (New York: Harper & Row, 1976), p. 10.

★ LBJ Plays Santa Claus ★

"The place: The Texas ranch of Senate Majority Leader Lyndon B. Johnson.

"The time: Several years ago.

"A young man arrives at the ranch to run a routine political errand. While he is waiting in the living room, Johnson strides in, picks up the telephone, and calls his press secretary, George Reedy, who is staying in the guesthouse down the road. He gives Reedy a monumental chewing out.

"'He was using language that I had never heard one human being use to another,' recalls the political worker, now an official of the University of Texas. 'He called him every filthy name in the book and some that aren't. I don't think I'll ever forget it.'

"Finally, Johnson hangs up the phone. He turns to the young man and says, 'Now, let's give George his Christmas present.' He leads the way outside to an expensive new station wagon, drives it down to the guesthouse, and toots merrily on the horn. When Reedy comes out, Johnson gives him the station wagon. To the astonished witness, the senator observes: 'You never want to give a man a present when he's feeling good. You want to do it when he's down.'"

SOURCE: James Deakin, "The Dark Side of LBJ," *Esquire*, August 1967, p. 45. Reprinted by permission of James Deakin.

★ A New Yorker Confesses ★

John V. Lindsay, mayor of New York City: "The only problem I ever have in New York City is people. People present me with a constant headache—we have too many slobs."

★ Advice on Immigration ★

When Congress was revising U.S. immigration laws in the late 1960s, Vice President Hubert Humphrey received this timely advice from an Indian living on a New Mexico reservation: "Be careful in revising those immigration laws of yours. We got careless with ours."

SOURCE: Bennett Cerf, *Laugh Day* (Garden City, N.Y.: Doubleday, 1965), p. 16.

★ EQUAL OPPORTUNITY ★

Virginia Military Institute has always prided itself on the fact that once a man passes through the gates of the school and becomes a cadet, his background, name, and circumstances no longer matter. The first blacks entered VMI in 1968. A reporter for a large metropolitan newspaper, somewhat dubious about the claim that blacks were being treated as equals at the school, asked one of the new black recruits about VMI life.

"This is the most equal place I ever heard of," came the reply. "Here they treat everybody like a nigger."

SOURCE: Henry A. Wise, *Drawing Out the Man: The V.M.I. Story* (Charlottesville: University of Virginia Press, 1978), p. 288.

★ NIXON COLLAGE ★

- At age twelve Richard Nixon told his mother, "I will be an old-fashioned kind of lawyer, a lawyer who can't be bought."
- As an anxious senior at Duke University Law School, Nixon broke into the office of the dean to find out his class standing. He discovered that he was at the top of his class. He was not punished.
- While touring Caracas, Venezuela, in 1958, Nixon was spat on by a protester. Secret Service agents grabbed the man and the then vice president planted a healthy kick in his shins. In *Six Crises*, Nixon recalled that "nothing I did all day made me feel better."
- Despite his boast that writing *Six Crises* was a maturing experience, Richard Nixon did not author the book. All of it was

ghosted by Alvin Moscow, except the final chapter on the 1960 campaign.

- Speaking to reporters about his health, President Nixon once claimed that he had never had a headache during his whole life.

★ NIXONISMS ★

- To an injured policeman waiting for an ambulance, Nixon remarked, "How do you like your job?"
- During the funeral of French president Georges Pompidou, Nixon declared: "This is a great day for France."
- One day a man was invited to the Oval Office to give President Nixon a chair made completely from a single piece of wood. The President sat in the chair, and it immediately collapsed. As he picked himself up from the floor, Nixon said nothing, but then asked, "Well, how do you go about doing this kind of work?"

SOURCE: Douglas Hallet, "A Low-Level Memoir of the Nixon White House," *New York Times Magazine*, October 20, 1974, p. 52.

★ EVEN THE MEDIOCRE DESERVE A ★ VOICE IN GOVERNMENT

Roman Hruska, speaking in the Senate in defense of Harold G. Carswell, Richard Nixon's nominee to the Supreme Court: "Even if he were mediocre, there are a lot of mediocre judges and people and lawyers. They are entitled to a little representation, aren't they, and a little chance? We can't have all Brandeises and Frankfurters and Cardozos and stuff like that there."

SOURCE: Richard Harris, *Decision* (New York: Ballantine, 1971), p. 117.

★ WATERGATE CONFUSION ★

Ron Ziegler, presidential spokesman during the Watergate crisis: "If my answers sound confusing, I think they are confusing because the questions are confusing, and the situation is confusing and I'm not in a position to clarify it."

SOURCE: Paul Morgan and Sue Scott, *The D.C. Dialect* (Washington: Washington News Books, 1975), p. v.

★ ROCKY WEARS BLACK ★

When Chiang Kai-shek died in April 1975, President Ford sent Vice President Nelson Rockefeller to Taiwan to attend the funeral. Shortly afterwards a reporter asked "Rocky," who had become vice president only a few months before, how Ford was going to use him. "It depends," replied Rockefeller, "on who dies."

SOURCE: John Lindsay, newspaper reporter, Washington, D.C., in a personal interview with Richard Shenkman at the Capitol, January 1979.

★ A MOMENT IN THE LIFE OF GERALD ★ FORD'S PRESS SECRETARY

Following is a remarkable exchange that occurred on Friday, September 3, 1976, between Ron Nessen, Gerald Ford's press secretary, and White House reporters on the subject of the President's campaign strategy:

Q: Ron, is the President going to make one speech every two weeks?

A: Probably a little more frequently than that. I think as Jim [Baker, campaign director] said he outlined the strategy pretty thoroughly at that time.

Q: Is that right, one a week?

A: Probably a little closer to two a week, yeah.

Q: Two a week.

A: No, I say one a week would be closer to the fact than one every two weeks.

Q: Ron, you said that this was a strategy reached after careful thought.

A: Um-hum.

Q: Could you share with us at this point what the thought was? What's he trying to accomplish by this strategy?

A: I think what I tried to say before, that the strategy is designed to elect the President. After all, the issue of this election is—You have a President who's been here for two years, who has a record, who has run up accomplishments, who has proposals for additional accomplishments. Do you want to keep him? Or, you have another candidate, with proposals and promises and positions on issues? Do you want to replace the President you have with the other fellow? Now that's the issue of the campaign.

Q: [Laughter]

A: You asked me a question. I'm answering it with our answer. If you want to hear it, okay, if you want to laugh it off I'll stop and go on to the next question. Okay, you asked me a question. I'm trying to answer it. This is the central issue of this campaign. Our strategy is designed to make the point that you should keep the President you have. There's a lot of people that have worked a lot of hours, including the President—and, as I say, you've all heard about it or seen the strategy book and so forth, and we are following our agreed-upon strategy.

Q: Is that the theme? Keep the only President you have?

A: It's one of the themes.

Q: That was one he used in Texas, where it didn't work too well. Is there any reconsideration—

A: Well, you know, I, okay.

SINCE WATERGATE

★ ≈≈≈

"I'm just confused. I don't know what I ought to do any-more. I don't know what works or what doesn't work. I don't know why this is happening. I'm just so confused."
—HILLARY RODHAM CLINTON

★ SCRAPBOOK OF THE TIMES ★

- When Jimmy Carter was asked in the 1960s to join the racist White Citizens Council, he said, "I've got $5 but I'd flush it down the toilet before I'd give it to you."
- Billy Carter was allergic to peanuts.
- When Jimmy Carter was president, he kissed the queen mother on the lips. She was appalled, as obituary writers noted upon her death.
- On the morning of the Wisconsin primary in 1980, Jimmy Carter held a press conference to announce that there had been substantial progress in talks to free the American hostages in Iran. Nothing came of this supposed break-through.
- In the 1990s, the Reagan Alumni Association, which is com-posed of 5,000 members of past Reagan campaigns, proposed replacing Alexander Hamilton's picture on the $10 bill with Reagan's.
- In the first Bush administration, when Secretary of Defense Dick Cheney complained one day that not one of his assis-tants supported Colin Powell's plan to deactivate thousands of small nuclear weapons in place in Europe, Powell, then chair-man of the Joint Chiefs of Staff, responded, "That's because they're all right-wing nuts like you."
- Ronald Reagan was considered by teachers to be a bright child. He could read by the age of five.

- When Ronald Reagan's father was told by a hotel clerk that the hotel did not allow Jewish guests, he stomped out and slept the night in his car.

- Jimmy Carter won over many Americans as an ex-president, but he infuriated two successors. On the eve of the 1991 Gulf War, Carter recommended in a letter to the members of the United Nations Security Council that they should oppose American policy. The Bush administration considered accusing him of treason. In 1994, during a crisis with North Korea over nuclear weapons research, Carter, in violation of instructions from the State Department, told North Korea the United States would agree to lift economic sanctions. A Clinton cabinet secretary exploded in rage that Carter was a "treasonous prick."

- Bill Clinton, as quoted by *Newsweek*: "I was a man of forty when I was sixteen and at forty a boy of sixteen."

- According to speechwriter Michael Waldman, Bill Clinton in 1993 reviewed every line of the federal budget—a first for any president.

- Bill Clinton loved to talk. His State of the Union address in 1995 was the longest any president ever delivered. He spoke for one hour and twenty minutes.

- The first time women were mentioned as a group in a presidential inaugural address was in 1997.

- On his last day in office, Bill Clinton granted a presidential pardon to Marc Rich, who was number six on the Department of Justice's list of "Most Wanted" international fugitives. Rich's ex-wife Denise had donated approximately $1.5 million to Clinton-related causes near the end of his term. Justice Department officials afterward noted that the pardon for Marc Rich was not processed through regular channels.

- Carlos Anibal Vignali, a drug dealer, and Almon Glenn Braswell, a convicted swindler, also were granted pardons in Clinton's final week in office. Hillary Rodham Clinton's brother, Hugh Rodham, lobbied to obtain the pardons and was paid $400,000 compensation. He later returned the money when ethical questions were raised about his fees.

- George W. Bush is related to two former presidents. His father, George H. W. Bush, was president. And W.'s fourth cousin, five times removed, was Franklin Pierce of Vermont, U.S. president from 1853 to 1857. Barbara Bush, whose maiden name is Pierce, is the only woman in U.S. history to be the wife of one president, mother to another, and the fourth cousin of another.

- An exact tally of the people killed in the September 11, 2001 terrorist attack on the World Trade Center in New York City is elusive. The New York Police Department lists 2,823, while the medical examiner's office lists 2,819. Other lists vary from 2,786 to 2,814. The death toll from the other attacks is undisputed: 224, not counting the hijackers; 184 at the Pentagon, plus 40 in Pennsylvania.

- On the first anniversary of the September 11 terrorist attack, the numbers that popped up for the New York Lottery were 9-1-1. "The numbers were picked in the standard random fashion," stated a lottery official. "It's just the way the numbers came up."

★ VIETNAM AFTER THE WAR ★

The war in Vietnam was one of the most divisive and unpopular wars in American history. The American military withdrew from Vietnam in 1973, and on April 30, 1975, North Vietnamese forces entered Saigon. The American ambassador and other officials escaped in a helicopter evacuation from the rooftop of the U.S. embassy. The American part of the war was over. What happened next in these countries continues to be of interest in America because of its past involvement and because of the large number of refugees who are now American citizens.

Following the communist takeover, from 500,000 to 1 million people were placed in reeducation camps. The total population of South Vietnam was about 20 million. The victims included not only former South Vietnamese soldiers but students, intellectuals, monks (both Buddhist and Catholic), and many political militants who had been in sympathy with the National Liberation

Front of South Vietnam (the communists) but quickly found themselves on the outs from their northern comrades.

Conditions in the reeducation camps varied considerably. While some camps were near towns and did not even have fences, other camps were in isolated jungle areas where seventy to eighty prisoners were placed in cells built for twenty. Hunger, disease, torture, and death were common. The typical "reeducation" stay in the camps lasted three to eight years. The last survivors of the reeducation programs did not return home until 1986. Of course, hundreds of thousands of others became "boat people," fleeing the country in anything that might float. Many of these refugees drowned or were killed by pirates. Today Vietnam remains one of the most oppressive nations on earth.

SOURCE: Stéphane Courtois et al., *The Black Book of Communism, Crimes, Terror, Repression* (Cambridge, Mass.: Harvard University Press, 1999), pp. 565–635.

★ RONALD REAGAN: TAX CUTTER? ★

In his first year in office, Ronald Reagan famously succeeded in winning congressional approval for the single largest tax cut in American history: a 25 percent cut in taxes across the board. That fixed in stone his reputation as a tax cutter.

But Reagan also increased taxes, though his doing so received far less attention. Over the course of his administration, Reagan raised taxes seven times. To disguise what he was doing, his handlers referred to most of these increases as "revenue enhancers."

One of the biggest budget challenges he faced was saving Social Security. To assure the program's continued solvency, Reagan approved what was, at the time, the largest tax increase in American history. This tax passed Congress less than two years after Reagan had signed the largest tax cut in American history. The tax hike brought in so much revenue that the surplus was used to help the government pay for other social programs for the next generation.

Taken together, Reagan's tax cuts were larger than his tax increases. But by the end of his term he had barely reduced the tax burden on Americans. According to Michael Kinsley, "Federal tax collections rose about a fifth in real terms under Reagan. As a share of [the gross domestic product], they declined from 19. 6 percent to 18.3 percent." Most Americans at the end of his term paid higher taxes than they did at the start.

★ WAS REAGAN SENILE AS PRESIDENT? ★

"In her book *Reporting Live*, former CBS White House correspondent Lesley Stahl wrote that she and other reporters suspected that Reagan was 'sinking into senility' years before he left office. She wrote that White House aides 'covered up his condition'—and journalists chose not to pursue it. Stahl described a particularly unsettling encounter with Reagan in the summer of 1986: her 'final meeting' with the president, typically a chance to ask a few parting questions for a 'going-away story.' But White House press secretary Larry Speakes made her promise not to ask anything. Although she'd covered Reagan for years, the glazed-eyed and fogged-up president 'didn't seem to know who I was,' wrote Stahl. For several moments as she talked to him in the Oval Office, a vacant Reagan barely seemed to realize anyone else was in the room. Meanwhile, Speakes was literally shouting instructions to the president, reminding him to give Stahl White House souvenirs. Panicking at the thought of having to report on that night's news that 'the president of the United States is a doddering space cadet,' Stahl was relieved that Reagan soon reemerged into alertness, recognized her, and chatted coherently with her husband, a screenwriter. 'I had come that close to reporting that Reagan was senile.'"

SOURCE: Jeff Cohen, "The Press Slept While Reagan Rambled" (www.TomPaine.com).

★ RONALD REAGAN ON MOUNT RUSHMORE ★

In 1997, disturbed by public indifference to Ronald Reagan, a group of supporters began a public campaign to raise his profile and improve his image. They called their campaign the Ronald

Reagan Legacy Project. It was wildly successful. In 1998 they succeeded in getting Congress to change the name of Washington National Airport to Ronald Reagan Washington National Airport. Soon after that they succeeded in getting a turnpike in Florida named after him, then a nuclear aircraft carrier. The group's goal? "We want to create a tangible legacy so that thirty or forty years from now, someone who may never have heard of Reagan will be forced to ask himself, 'Who was this man to have so many things named after him?'" explained Michael Kamburowski, the Reagan Legacy Project's executive director. William Buckley lobbied to get Reagan placed on Mount Rushmore.

★ ANATOMY OF A POLITICAL HIT ★

Politicians never know for sure which story or event will take hold in the public mind. What words or actions will be a defining moment? Will the event even be true?

On February 5, 1992, George H. W. Bush was running for reelection. On that day of campaigning, Bush stopped by the National Grocers Association convention in Orlando, Florida. One of the exhibits Bush visited displayed a new type of supermarket scanner that could not only price groceries, but could weigh them and read mangled or torn bar codes. Gregg McDonald of the Houston *Chronicle* was the only newspaper reporter who accompanied Bush, and he later filed a two-paragraph story for the press pool that stated Bush had a "look of wonder" on his face when shown the latest in grocery store technology. McDonald's full story for the *Chronicle* that day did not mention Bush had viewed any grocery store scanner equipment.

The next day, the New York *Times* ran a front-page story by Andrew Rosenthal that claimed Bush had "emerged from 11 years in Washington's choicest executive mansions to confront the modern supermarket." The President, according to Rosenthal, "grabbed a quart of milk, a light bulb, and a bag of candy and ran them over an electronic scanner. The look of wonder flickered across his face. . . . 'Is this for checking out?'" The story then continues with Bush saying, "I just took a tour through the exhibits here. Amazed by the technology." The front-page story continued by reminding readers that supermarket scanners had been developed in 1976 and had been in general use for over a decade. The impression that Bush was out of touch with ordinary people was unmistakable. Of course, the Rosenthal piece failed to mention that Rosenthal himself was not there when all of this "occurred," and that the entire basis for his article was fabricated from McDonald's mention in the pool report of a "look of wonder."

Newspapers and especially editorial writers around the country quickly picked up and repeated the *Times* story. Here, in one concise anecdote, was a larger truth about cloistered career politicians. The next day the Boston *Globe* ran a story called "President Bush Gets in Line." The Washington *Post* followed with "Grocery Shopping Needn't Concern President." Anthony Lewis of the New York *Times* editorial page piled on further with "The Two Nations."

Within a week, however, other reporters realized the Rosenthal piece was being taken seriously and that, at best, it was a very questionable interpretation of what actually had happened. Stories began to appear in other newspapers criticizing the *Times*'s methodology. *USA Today* ran a story on page two called "Bush: Scanner Story Was Short-Change Job." The Atlanta *Journal-Constitution* followed with, "How President Bush Got Dunked by the Pool." The Washington *Post* ran, "The Story That Just Won't Check Out."

Now with egg on its face, the New York *Times* found a videotape of the event and ran a story proclaiming that Bush had reviewed both ordinary and newfangled scanners, and was unfamiliar with and impressed by both. But no one else who viewed the video saw it that way. *Newsweek* reported that Bush was "curious and polite, but hardly amazed." *Time* stated that the exchange was "completely insignificant as a news event. . . . If anything, he was bored." And Bob Graham, the salesman who had demonstrated the scanner technology for Bush, stated, "It's foolish to think the president doesn't know anything about grocery stores. He knew exactly what I was talking about."

The story faded quickly; the New York *Times* had its opinion of events, and everyone else who checked into it had another. But the perception lingered. Public attention had shifted away from the Gulf War and was beginning to focus on the not-so-great economy. President Bush had first said a recession would not happen; then when it did, he promised the recession would end soon. He seemed older and out of touch, someone who would no more know how to handle the troubled economy than he would know the price of a quart of milk. His lead in the polls began to slip away. And, after all, didn't we hear somewhere that the president doesn't even know that grocery stores have electronic scanners?

SOURCE: Barbara and David Mikkelson, *Maybe I'm Amazed: Urban Legends Reference Pages* (www.Snopes.com, 2001).

★ CLINTON-ERA QUOTES ★

- Bill Clinton's advice to Gennifer Flowers, which she taped, on what to say if she was ever asked about a possible affair with Clinton: "Deny that we ever had an affair."
- Bill Clinton, on *60 Minutes*, after denying having an affair with Gennifer Flowers: "I have absolutely leveled with the American people." (Same interview: "I have acknowledged that my marriage is not perfect.")
- Bill Clinton, referring to Monica Lewinsky: "Listen to me. I never had sexual relations with that woman."
- Bill Clinton, testifying in the lawsuit filed by Paula Jones: "That depends on what the meaning of the word *is* is."
- Bill Clinton, in the brief speech to the country acknowledging he had had an affair with Monica Lewinsky: "I misled even my family."
- Bob Kerrey, Democratic senator from Nebraska: "Clinton's an unusually good liar. Unusually good. Do you realize that?"
- Johnny Chung, the former Democratic Party fund-raiser who told federal investigators that China's chief of military intelligence funneled $300,000 through him to the 1996 Clinton reelection campaign, in an interview with the Los Angeles *Times* (July 1997): "I see the White House is like a subway: You have to put in coins to open the gates."

- Top adviser James Carville on Paula Jones, who claimed Clinton had sexually harassed her: "Drag $100 bills through trailer parks, there's no telling what you'll find."
- Bill Clinton, speaking to a group of wealthy contributors, some of whom remained angry at Clinton for raising their taxes in 1993: "It might surprise you to know that I think I raised them too much too."
- Webb Hubbell, former senior Department of Justice official and longtime Clinton friend, in the course of a conversation with his wife from prison: "So I have to roll over one more time."
- In 1992, Bill Clinton promised during the campaign for the presidency to establish "the most ethical administration in the history of the republic."

★ THE CLINTON MYSTERIES ★

Three great mysteries provoked endless debate during the Clinton years. Perhaps the most puzzling of all concerned the Paula Jones lawsuit. Jones claimed that Clinton, when he was governor of Arkansas, had exposed himself to her in the privacy of a hotel room. After he became president she filed a sexual harassment lawsuit against him in federal court. Jones's former attorney told NBC's *Dateline* that on two occasions Clinton was given the opportunity of ending the suit merely by making an apology and a $25,000 payment. Yet even after winning reelection in 1996, he refused. Jones continued to depose anyone and everyone she thought might be helpful to her case, eventually including Monica Lewinsky, Kathleen Wiley, Juanita Broaddrick, and many others. Clinton's obtuse answers (or lies) to a federal court in the Jones case eventually precipitated the filing of impeachment charges against him.

A second great mystery is Bill Clinton's affair with Monica Lewinsky. Why had he been so reckless? The affair began when Lewinsky, a young White House intern, delivered a pizza to Clinton in the winter of 1995 when the government had been shut down as a result of a high-stakes political dispute between the Republican-controlled Congress and the Clinton White House. At the time, Kenneth Starr, the "special prosecutor" appointed to investigate a failed real-estate transaction known as Whitewater, was searching the President's past for damaging evidence. Paula Jones's lawyers were looking for evidence to strengthen her case. Clinton, knowing that both Starr and Jones were after him, nonetheless carried on an affair with Monica Lewinsky. Among presidential mysteries this surely must rank as one of the most mysterious of all.

A third mystery was the public's continued enthusiasm for Bill Clinton throughout the Lewinsky scandal. In all of American history there is no other parallel circumstance. At first, when the scandal broke on the Web site of Matt Drudge, pundits speculated that it might ultimately cost Clinton his presidency. *ABC News* reporter Sam Donaldson predicted that Clinton would be forced out of office within a matter of days if the allegations proved true. Dick Morris, the President's pollster, conducted a survey to find out how the public was reacting to the news. Morris told Clinton that the public would tolerate infidelity but not lying about it. Clinton, in fact, lied and lied about the affair, denying to his wife, his friends, his cabinet, and the country that he had had "sex with that woman, Ms. Lewinsky." And yet in the end the public not only tolerated their president lying, they sent his poll numbers soaring.

Some have explained that Clinton's numbers remained high because he dragged out the scandal. Others note that Kenneth Starr was such a perfect foil that Clinton was able to rally support simply by opposing Starr. Though the House of Representatives impeached him, by the time the case reached the Senate there no longer was any drama left in the scandal. Clinton would escape conviction and save his presidency.

A final mystery from the Clinton administration began in 1993, just six months into his presidency, when White House aide Vince Foster was found dead in a wooded section near the George Washington Parkway. From the first the White House believed it was a suicide. Within days the park police, who initially handled the investigation, came to the same preliminary conclusion. But both Clinton's enemies and friends agreed—though for different reasons—that there was an air of mystery surrounding Foster's death. White House officials could not understand why Foster had killed himself, George Stephanopoulos telling the Washington *Post*, "The fundamental truth is that no one can know what drives a person to do something like this. Since you can't ever know, it's impossible to speculate on it. In the end, it is a mystery." Adding to the doubts was the "belated discovery" of a torn note at the bottom of Foster's briefcase, leading to speculation that White House aides had improperly tampered with evidence and perhaps concealed something embarrassing to the administration. Foster's death was officially ruled a suicide. But so many questions were raised by Clinton's enemies that doubts long persisted about the official story.

★ BILL CLINTON'S TELEPROMPTER PROBLEM ★

On September 22, 1993, Bill Clinton spoke before a joint session of Congress on behalf of his plan to revolutionize the health care system. When he stepped up to speak he noticed that technicians had loaded the wrong speech in the TelePrompTer. For ten minutes he spoke from memory while aides scrambled to install the correct speech.

Then it happened again. Almost. In 1997, as speechwriter Michael Waldman was riding in a limousine from the White House to Capitol Hill, he decided to put some final touches on the State of the Union address the president was about to deliver. Waldman, working on the speech on his laptop, added a single comma. Somehow doing so removed all of the formatting from the document. Suddenly, all of the text ran together, making it nearly impossible to read. As Clinton was making his way to the podium, Waldman worked behind the scenes with the Marine aide in charge of the TelePrompTer machine to enter paragraph breaks by hand. They got it done just as Clinton began to speak.

SOURCE: Michael Waldman, *POTUS Speaks* (New York: Simon & Schuster, 2002).

★ PEARL HARBOR: THE MOVIE ★

In 2001, on the sixtieth anniversary of Pearl Harbor, CNN aired a story about the just-released Disney movie about the Japanese attack. In the middle of the story, up popped some black-and-white footage of gunners firing their weapons into the sky. According to the CNN reporter, this was "actual footage" of the Pearl Harbor attack. That is, this was not like the footage in the rest of the story that came from the Disney movie. This was real!

Viewers who were paying attention might have wondered. The shots didn't look like they came from a home movie. They were perfectly framed. You could almost swear a director was off-camera shouting directions to the photographer to be sure to make the gunners look as heroic as possible. About the only difference between these pictures and those from the Disney movie was that Disney's were in color.

How did the photographer manage to get such picture-perfect shots amidst the chaos of the Pearl Harbor attack? By shooting them on a Hollywood back lot, that's how. They come from the movie *December 7th*, directed by John Ford. A short while after Pearl Harbor, Ford re-created in Los Angeles the Japanese sneak attack. He had to re-create the scene because there were hardly any pictures from the real attack. According to film historian Rick Decroix, there exists less than one minute's worth of actual footage. All the rest of the footage you've ever seen was shot by Hollywood.

CNN's story was not the first to mistakenly pass off Ford's movie footage as the real thing. Decroix says he knows of at least two documentaries that did so as well a decade earlier at the time of the fiftieth anniversary of Pearl Harbor.

Blame the mix-up on the federal government, which sponsored the movie. To inflame public opinion against the Japanese, American officials wanted the public to believe they were seeing actual battle footage. So CNN and the others were taken in by a government propaganda plot hatched decades ago during World War II.

★ ALEXANDER GRAHAM WHO? ★

On June 11, 2002, to little fanfare, the U.S. House of Representatives declared that the telephone was invented by an Italian-American named Antonio Meucci, a sausage and candle maker. Forget Alexander Graham Bell. The House declared that Bell's patent for the telephone was based on "fraud and misrepresentation."

News of the House resolution was slow to circulate. When the media contacted the curator of the Bell Homestead Museum in Brantford, Ontario, he said he was surprised. He hadn't heard of the resolution. In Italy the news was greeted warmly, an Italian paper referring to "Bell as an impostor, profiteer, and a 'cunning Scotsman' who usurped Meucci's spot in history, while Meucci died poor and unrecognized."

Is it true that Meucci, not Bell, invented the telephone? Robert Bruce, the Pulitzer Prize–winning biographer of Bell, tersely dispatched the Meucci claim. "It's ridiculous," he said.

Meucci claimed that "by means of some little experiments, I came to discover that with an instrument placed at the ear and with the aid of electricity and a metallic wire, the exact word could be transmitted holding the conductor in the mouth." Bruce says he was deluded. Meucci's patent, says Bruce, was "essentially the same as connecting two tin cans with a string."

Italian-Americans have long claimed that Meucci had been cheated of the honor of being recognized as the telephone's inventor. Only one historian, however, took his claims seriously, Giovanni E. Schiavo, in a book published in 1958. Bruce says that Meucci not only failed to invent the telephone, he "did not understand the basic principles of the telephone either before or after Bell's invention."

The resolution honoring Meucci was introduced by Staten Island representative Vito Fossella. Fossella, claiming he based the resolution "on our study of historical records," said he pressed for its passage "to honor the life and achievements long overdue of Antonio Meucci, a great Italian American and a former great Staten Islander." The House allotted forty minutes to debate the measure. Five members of Congress spoke in favor; none spoke against. The resolution was approved by voice vote: "*Resolved,* That it is the sense of the House of Representatives that the life and achievements of Antonio Meucci should be recognized, and his work in the invention of the telephone should be acknowledged."

SOURCE: History News Network (www.hnn.us/articles/802.html).

★ BUSH-SPEAK ★
QUOTATIONS FROM PRESIDENT GEORGE W. BUSH

- "I promise you I will listen to what has been said here, even though I wasn't here."—speaking at the President's Economic Forum in Waco, Texas, August 13, 2002

- "The problem with the French is that they don't have a word for entrepreneur."—discussing with British prime minister Tony Blair the decline of the French economy

- "The public education system in America is one of the most important foundations of our democracy. After all, it is where children from all over America learn to be responsible citizens, and learn to have the skills necessary to take advantage of our fantastic opportunistic society."—May 1, 2002

- "And so, in my State of the—my State of the Union—or State—my speech to the nation, whatever you want to call it, speech to the nation—I asked Americans to give 4,000 years—4,000 hours over the next—the rest of your life—of service to America. That's what I asked—4,000 hours."—Bridgeport, Connecticut, April 9, 2002

- "I understand that the unrest in the Middle East creates unrest throughout the region."—Washington, D.C., March 13, 2002

- "There's nothing more deep than recognizing Israel's right to exist. That's the most deep thought of all. . . . I can't think of anything more deep than that right."—Washington, D.C., March 13, 2002

- "My trip to Asia begins here in Japan for an important reason. It begins here because for a century and a half now, America and Japan have formed one of the great and enduring alliances of modern times. From that alliance has come an era of peace in the Pacific."—Tokyo, February 18, 2002 (Bush's father, George H. W. Bush, a bomber pilot during World War II, was shot down by the Japanese in 1945 and escaped death only by a daring open-sea rescue.)

- "The legislature's job is to write law. It's the executive branch's job to interpret law."—Austin, Texas, November 22, 2000

- "They want the federal government controlling Social Security like it's some kind of federal program."—November 2, 2000

- "I know what I believe. I will continue to articulate what I believe and what I believe—I believe what I believe is right."—Rome, July 22, 2001

SOURCE: www.politicalhumor.about.com/library/blbushisms.htm.

★ THE FIRST AMERICAN? ★

About 9,200 years ago, a man lived and died by the northern shore of the Columbia River near present-day Kennewick, Washington. He was a tough guy. Years before his death a stone spear point had been hurled into his hip and never removed. Later his chest had been crushed, and his left arm had withered. Yet he lived to be about 45 years old, and he probably died from an infection. And he apparently was not a "Native American," since his features did not match those of the present-day Indian people of the Pacific Northwest. He is one of North America's most exciting archaeological finds and a big political problem.

Nine thousand years later, give or take a few years, two young men waded into a shallow area of the Columbia to get a better view of the final race at the 1996 annual Tri-Cities Water Follies hydroplane races. One of the men stepped on a round object, put his hands in the water to inspect, and pulled out a human skull. Being true sports fans, the men hid the skull near a bush and returned to their spot to view the final hydro race. They then called the Kennewick police, who called local anthropologists. About 90 percent of the man's skeleton was recovered from the shallow water. At first police and anthropologists assumed the man was an early European settler from maybe a hundred years ago. But the spear tip lodged in his hip was obviously older, so they sent a small bone from the hand to the University of California, Riverside, to be carbon dated. The results were shocking. Kennewick Man, as the skeleton came to be called, had lived and died in southern Washington some 9,200 to 9,600 years ago. He was literally older than Moses.

So who was Kennewick Man? The few anthropologists who examined him reported he did not resemble a modern American

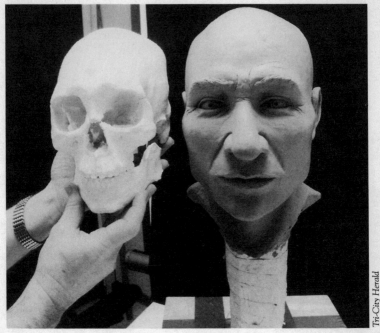

Tri-City Herald

The skull of Kennewick Man, the 9,200-year-old "First American"
recently found in a shallow area of the Columbia River near Kennewick,
Washington. This skull has caused some to speculate on the possibility
that the earliest Native Americans may have looked more like Captain
Picard from *Star Trek* than any present-day Native Americans.

Indian. He was tall and thin with a long skull. His face was narrow
and prognathous, not broad and flat. "I thought we had found an
early European settler," speculated one expert after seeing Kennewick Man's Caucasian features. A white guy living in the New
World 9,200 years ago? No one could know Kennewick Man's skin
color. The only thing for certain was that his remains showed that
the original settlement of the Americas was more complicated
than previously imagined.

But Kennewick Man had been unearthed into our modern,
political world. Because Kennewick Man was found on federal

property, the Army Corps of Engineers assumed possession of the body. Almost immediately upon discovering Kennewick Man's age, five Pacific Northwest Indian tribes claimed his body under the Native American Graves and Repatriation Act, a law designed to return American Indian remains to tribes. The government immediately agreed. The tribes planned to rebury Kennewick Man in a traditional tribal ceremony and not let any scientist study the skeleton. "Let the anthropologists study their own bones," stated one Yakima tribe official.

Eight scientists, including two from the Smithsonian Institution, filed a lawsuit against the government and the tribes to allow scientific examination of the skeleton. Whether Kennewick Man was an ancient relative of these tribes was very debatable, they argued. On preliminary examination he did not look anything like a member of these tribes, or any other traditional "Native American." Study of the skeleton would be valuable in understanding ancient migration patterns into the New World. The case would drag through the court system for years.

Both the Indians and the government had an interest in claiming Kennewick Man was Indian. The tribes wanted to preserve the impression that they are the original Native Americans. The government wanted to maintain good relations with the tribes. Indeed, an internal memo from the Army Corps of Engineers, which routinely negotiated with the tribes over fishing and water rights, stated, "All risk to us seems to be associated with not repatriating the remains."

And what of the site where Kennewick Man was found? Was it an ancient burial ground or village? Or had Kennewick Man been alone when he died? This is unknown—and now never will be known. After initially allowing some preliminary archaeological work to proceed, the Corps of Engineers suddenly reversed itself and announced that the site had to be destroyed to help control erosion along the Columbia River bank. Doc Hastings, a congressman from Washington who had lived in an area overlooking the site, disagreed. "There is no significant erosion," Hastings explained. "That part of the river isn't even free flowing—it's part of a lake behind a dam." With Hastings's help, both the U.S.

House of Representatives and the Senate passed bills to halt the planned destruction, but before the two bills could be reconciled, the Corps of Engineers dumped 500 tons of rock and gravel from helicopters on top of the area. An additional 300 tons of dirt and logs were moved in to layer the river shoreline, and thousands of trees were planted on the remade terrain. Archaeologically speaking, the site had been destroyed. Other Corps documents indicated a "concern on the part of the [Clinton] White House" led to the quickly executed erosion-control plan.

In September 1998, a federal judge ordered Kennewick Man moved to Seattle's Burke Museum for safekeeping. The Corps of Engineers had not done a good job of babysitting the ancient skeleton. While the body was off limits to scientists and the public, the Corps had lent him out four times to Indian groups for ceremonies. When the Burke finally took an inventory of the ancient bones, major pieces were missing. The stolen bones have never been recovered.

The central question before the court was, What constitutes a "Native American"? The Native American Graves and Repatriation Act clearly states that any skeleton that can be linked to a specific modern tribe shall be turned over to that tribe. The tribe may or may not study the skeleton "scientifically." But what about ancient skeletons like Kennewick Man, which may lack any connection to a modern tribe? Indian activists and the federal government claimed that every skeleton dated from before Columbus in 1492 is obviously "Native American" and must be remitted to some tribe, even if no direct link can be found between the skeleton and the tribe. Scientists claimed that the Graves and Repatriation Act applied only to bones linked to a modern people. Truly ancient bones should be the property of all Americans, available for study to help explain our history. The question of the status of ancient bones hovers over more remains than just those of Kennewick Man. Nevada's 9,400-year-old Spirit Cave Mummy, Idaho's 11,000-year-old Buhl Woman, and a recently uncovered 11,000-year-old teenage girl in Brazoria County, Texas, are all caught in a similar tug-of-war between anthropologists and modern Indian tribes.

In October 1999, over three years after Kennewick Man's discovery, two federally appointed scientists issued the most complete report on Kennewick Man to date. The court had ordered this study to help determine if Kennewick Man could be linked to a modern tribe. Not surprisingly after 9,200 years, no link could be established. The study said Kennewick Man is not like any modern Indian or European, but most likely came from southern Asia. His closest modern relatives appear to be from Polynesia or the Ainu of Japan. DNA testing proved unsuccessful because of the age of the bones. The report, while welcomed by anthropologists, was well short of a full, refereed anthropological study.

Finally, on August 30, 2002, six years after the discovery, federal judge John Jelderks ruled that because Kennewick Man has no clear connection to any present-day tribes, he could be studied by scientists. The judge repeatedly criticized the Corps of Engineers and the Department of the Interior for their handling of the case. Four Pacific Northwest Indian tribes, the Colville, Yakama, Umatilla, and Nez Perce, appealed the ruling to the 9th Circuit Court. The federal government joined the appeal on the side of the Indians and against the scientists. Both sides claimed the matter may eventually have to be settled by the Supreme Court.

SOURCES: James C. Chatters, *Ancient Encounters: Kennewick Man and the First Americans* (New York: Simon & Schuster, 2001); *Tri-City Herald* (Kennewick, Pasco, and Richland, Washington); Web site: www.Kennewick-Man.com.

INDEX